Progress in Construction Science and Technology

Edited by:

Roger A. Burgess, BArch, ARIBA, FIOB, FIWSP
Peter J. Horrobin, MA, ARIC
John W. Simpson, BSc, MSc, AInstP

Senior Lecturer and Lecturers respectively in the Department of Building of the University of Manchester Institute of Science and Technology

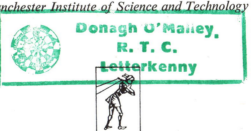

ASIA PUBLISHING HOUSE

Bombay—Calcutta—New Delhi—Madras—Lucknow
Bangalore—London—New York

Published by
Medical and Technical Publishing Co Ltd
Chiltern House,
Aylesbury, Bucks

SBN 852 00015 4

First published 1971

A/690

PRINTED AND BOUND IN GREAT BRITAIN BY
HAZELL WATSON AND VINEY LTD
AYLESBURY, BUCKS

Progress in Construction Science and Technology

Preface

The aim of this book, the first of a series of such volumes, is to provide a vehicle for the publication of comprehensive review articles on aspects of construction science and technology which have application in the many divisions of the construction industry.

As a glance at the contents page will confirm, the present volume contains reviews of a wide variety of important and relevant topics, written by an international and distinguished team of contributors.

There are many inadequately documented topics within the field of construction science and technology which could be very usefully reviewed in a volume of this type. Consequently we will be pleased to consider ideas for future papers from potential contributors. It is our hope that in the second volume we will also be able to publish short reports of progress made in fields not broad enough for a comprehensive review.

Without the help and co-operation of our colleagues in the Department of Building of the University of Manchester Institute of Science and Technology this book would probably have never been published and we would like to acknowledge here our gratitude to them.

ROGER BURGESS
PETER HORROBIN
JOHN SIMPSON

Manchester, December 1970

Contents

1
Sound insulation of partitions

H. Derrick Parbrook, B Sc, Ph D
Professor of Building Science,
University of Liverpool
Kenneth A. Mulholland, B Sc, M Sc (Eng), Ph D
Lecturer in Building Science,
University of Liverpool

1.1. INTRODUCTION

This article is a review of the present knowledge of the airborne sound insulation of partitions. A discussion of the main experimental findings is followed by a survey of the theory and the problem of measurement is briefly reviewed in an appendix: an understanding of basic concepts and terminology is assumed. The authors are very conscious that to attempt a review of such a wide field as this in a few pages requires a rigid selection of topics but they hope that they have largely chosen to discuss mainstream work.

It is useful to see the subject in context. The fabric of a building performs many functions; one of these is to limit, to values acceptable to the user, the amount of sound and vibration energy entering a room from sources in neighbouring rooms and outside the building. A quantitative knowledge of the values acceptable to the user and the means of predicting the acoustic performance of a given construction to the required accuracy are essential elements in a logical design process. Two approaches to this have been used.

One approach is to specify numerically the noise level that is acceptable in a particular room, bearing in mind the use and position of the room. The noise level surrounding the room must be measured or predicted. The difference between these figures represents the amount of insulation required. From theory and/or experience, composite constructions are chosen to meet the requirements. There are many problems in this approach:

a. Criteria for acceptable sound levels are based on a consensus concerning the human reaction to the aural climate. They are quantities attempting to express numerically an average reaction to an average stimulus in an average environment. Most criteria can only be tentative.

b. There are difficulties in specifying measurable quantities that correlate with subjectively acceptable aural conditions. Several quantities are used; of these the dBA system for steady state sounds is simple, but of limited accuracy. The NC (1) and ISON (2) systems are more accurate and of greater usefulness, but they only apply to a limited range of noises. Additional factors which can be taken into account with these simple systems include the environment of the accommodation, the time pattern of the sound, the noise history of the occupants, and the quality of the intruding and ambient sounds [see, for example, BS 4142 (3)].

An alternative approach is to find constructions in the field which lead to acceptable aural conditions for most people and list these as 'deemed to satisfy' constructions—as in the English Building Regulations (4). It is possible to measure in the field—using standardized techniques—the insulation given by constructions known to give acceptable aural conditions for most people and publish the measured insulation figures as a criterion [see, for example, the Scottish Building Regulations (5)]. Allowed deviations are also specified. Somewhat similar approaches to the Scottish Building Regulations are used in several countries and these have also been generalized to rate constructions other than those for domestic accommodation [for example, see ISO R717: 1968 (6)]. There are also difficulties with this approach:

    a. The English Building Regulations refer to houses and flats and define only party walls (and floors) that lead to acceptability. The insulation of a total construction can depend markedly on the flanking transmission, the area of the partition, the absorption present in the room and the stiffening effects of such elements as chimney constructions in the partition. Moreover, there is some evidence that the definition of deemed-to-satisfy party wall constructions has an inhibiting influence on the development of new constructions. It is assumed with no great justification that a deemed-to-satisfy construction will have the same acceptable performance when erected by different contractors.

    b. There are confusing differences in the presentation of results and recommended procedures for acceptable constructions.

Both of these approaches to design and construction assume that measurements can be made that are easily reproducible, are significant, and generally acceptable. There are problems in measurement and these are discussed in the Appendix.

Ideally we should be able to predict in a simple manner the performance of a composite construction in the field from a knowledge of those basic physical properties of the individual elements of the construction, such as superficial mass, stiffness, and porosity, that are easily measurable in a simple laboratory. This is a long-term objective and much work remains to be done. In the short term we need to have national and international agreement on a number of matters. These include:

a. The need to base our concepts on measurements in the field rather than in the laboratory until we quantitatively understand the differences which exist between these two measurements; the field constructions measured should be specified in an agreed manner.

b. Simple corrections necessary for the area of a partition, the volume and reverberation time of the room, the flanking transmission, and for the differences of the sound field when these deviate in a minor (and specified) way from those holding when measurements were made on the basic construction.

c. Where appropriate, the method of rating the acceptability of constructions for various uses.

1.2. MEASUREMENT DATA

There is a large and sometimes confusing literature on the sound insulation of panels. In examining this work the varying conditions under which measurements were made should be noted. These can vary from field measurements in occupied accommodation to those in accommodation still in the finishing stages of construction, from measurements in transmission laboratories built to varying specifications to those using a simple transmission loss box. Various techniques of measurement and normalization are used. The deviations from (true) random incidence of sound energy on the panel can be marked; some results are quoted for normal incidence, some for claimed random incidence, and some for an angle of incidence of 45° which are stated to approximate closely to random incidence (7). The edge mounting conditions are rarely specified with precision. There are also uncertainties in the repeatability of measured results.

This is especially noticeable at the lower frequencies where the results are not only a function of the panel performance but also apparently of the size and shape of the testing facility rooms. Utley (8) has found that laboratory measurements on a given panel construction can vary by $\pm 1 \cdot 5$ dB above 315 Hz and by considerably more below. Measurements on approximately 6·25 mm glass show a 20 dB range at 100 Hz and 4 dB at 500 Hz. At this latter frequency one might expect that standing wave and resonance phenomena would have small effect and that a repeatability of 1 dB would be readily achieved. Minor difficulties in summarizing the published work arise from the use of Metric and Imperial units by different workers, from varying objectives of the experimental work and lack of quoted tolerances. On this last point as a non-trivial example one can ask the precise significance of the three adjectives in the phrase, 'a 9-inch plastered brick wall'.

A number of collected sets of sound insulation results are available in the literature (9–11).

1.2.1. Single panels

Mass effects. For single, non-porous, homogeneous panels measurements show that the insulation (transmission loss—TL) at a given frequency and also the mean insulation (the average of the 16 one-third octave measurements between 100 and 3150 Hz) are both functions of the mass per unit area. Cook and Chrzanowski (12) report a relation $TL = 13 + 14 \cdot 5 \log_{10} m$ (m is the superficial mass of the panel in kg/m²). Pilkington Brothers Ltd (13) quote a value for glass of $TL = 19 \cdot 34 + 10 \cdot 5 \log_{10} d$ with a correlation coefficient of 0·84 (d is the glass thickness in mm). This suggests that there is an increase of 3–4·4 dB for each doubling of the superficial mass. However, one must have grave doubts about the meaning of TL, for many panels show prominent coincidence effects within the measurement range; an increase in m increases d and lowers the critical frequency.

Coincidence effects. The coincidence effect (14) is well established. For most homogeneous materials the agreement between the calculated critical frequency [using Cremer's infinite panel theory (14)] and the experimental values is good. Some disagreements may be attributed to the use of bands of frequencies for measurements or to varying angles of incidence of the energy on the panel. The reduction in the insulation below the mass law value at and near the coincidence frequency has only recently been predicted by Crocker (15) using modal theories and finite-sized panels. For 6·4 mm glass at 2000 Hz, a reduction of 5–12 dB below the 1000 Hz value is reported (16). For materials such as 12·7 mm plasterboard and 12·7 mm asbestos–cement sheets reductions of 15 dB are found. 9 in (230 mm) brick-work has a coincidence frequency below 100 Hz which lies outside the normal frequency range for measurement.

Edge-mounting effects. Flanking transmission is normally understood to arise from the flow of sound energy from a panel through the edges of the panel and the surrounding supporting framework to other panels or systems where it is radiated into the neighbouring space. Experiments have been carried out to determine the effect of such transmission. Kurtze, Tamm, and Vogel (17) examined the transmission of bending waves around a 90° bend. They found that transmission was dependent on the type of joint used at the bend: a rigid joint gave an attenuation of 3–5 dB, a rolling joint, 12 dB; a rubber interlayer gave an attenuation of 20 dB for a 2 mm layer and 30 dB for a 30 mm layer. Westphal (18) investigated a cross and tee junction and found that vibrations travelling through a cross were attenuated by up to 10 dB at 1 kHz and through a tee junction by 7 dB between the vertical and each of the right-angle members.

Kurtze (19) has also reported on the transmission of bending waves at junctions with finite L-shaped branches and with constrictions and varying thicknesses. Resonance effects were also observed. Sharp *et al.* (20) examined the reduction of flanking transmission when a soft rubber surround was placed between the edges of a steel plate and its supporting framework. He found that the increase in insulation was negligible except at coincidence where transmission loss was increased by 3 dB. Utley's work (21) and some work of Eisenberg reported by Hochbrugge (22) support this view. There appears to be no evidence to support or deny the idea that other forms of panel resonance might also cause flanking transmission.

Apertures in panels. The simplest assumption concerning the transmission of sound through an aperture in a panel is that the sound energy transmitted by the aperture is equal to the amount of sound energy falling on the face of the aperture. This is called the 'area effect'. Gomperts and Kihlman (23) have reported that small cracks transmit more sound energy than this. Ingerslev and Neilsen (24) supported the 'area effect' with the provision that it only works when the diameter of the aperture is greater than the thickness of the panel. Wilson and Soroka (25) have shown that apertures and slits having finite depth can have resonances within themselves causing greater loss of insulation than would be expected by the area effect.

Mulholland and Parbrook (26) have demonstrated that the transmission of sound through a thin aperture is equal to that predicted by the area effect providing that the aperture diameter is greater than the wavelength. When the diameter is less than the wavelength the transmission coefficient of a thin hole tends to 1·2. They also claim that the transmission coefficient of individual members of an array of apertures is much increased at low frequencies, an experimental value of up to 5 being reported.

There is little doubt that the effect of edge cracks on the sound insulation of panels is critical. Eisenberg (27) points out that the insulation of a glass panel is independent of the mass of the window over a considerable range of masses when the window is unsealed and has opening elements. Aston (28) agreed with this result and showed improvements in insulation of 4–8 dB in the insulation if they are sealed with plasticine; even greater improvements have been reported by Pilon (29). The type of sealing used does not appear to be critical; Woolley (30) has reported that a difference of 1 dB was obtained between the improvements gained from sealing using tape and using putty.

Damping layers can be incorporated into simple panels in two ways. The layer can be placed over the surface of the panel (an unconstrained layer) or it can be sandwiched between two non-damping layers (a constrained damping layer). With the unconstrained layer damping arises from the stretching of the layer as the plate bends beneath it. Work on such layers has been reported by Oberst (31), Oberst and Becker (32), Lienard (33), and Kurtze and Watters (34). In the constrained case, the damping arises from shear motion of the damping layer between the two plates. Often such a treatment consists of a layer of damping material on a plate with a foil on its other surface; the symmetrical sandwich plate is a special case of this treatment.

Kerwin (35) and Moss, Kerwin, and Dyer (36) report on the behaviour of such a constrained multilayer panel. The two methods of
damping have equal effect when the weight of the damping layer is
between 10 and 20 per cent of the weight of the panel. Below 10 per
cent the constrained method is more effective, above 20 per cent the
unconstrained layer is the most effective.

Libby *et al.* (37) have shown that a 1·14 mm intermediate layer
gives more damping than a 0·38 or 0·76 mm interlayer but that no
further improvement is achieved by using an interlayer 1·52 mm
thick. The results suggest that the effect of damping on transmission
loss mainly controls the loss of insulation caused by the coincidence
effect. Measurements made at various angles of incidence (27) would
appear to support this. It is also true that laminating provides a means
of increasing the thickness and the mass of the over-all panel without
lowering the coincidence critical frequency. Each individual non-
damping layer retains its own critical frequency and the damping
constrains the composite panel from exhibiting its own coincidence
effects. In the case of low damping, this effect has been shown to
breakdown and double coincidence dips have been measured by
Cummings (38).

Panel size. There is disagreement in the reported experimental findings. Certainly until recently, much theoretical work ignored the effect
of panel size. Measured values by Eisenberg (27) have shown that the
effect of pane size in windows is relatively unimportant compared
with other factors. Low-frequency sounds could cause vibration in
the fundamental mode of the window but apart from this, he does not
report any significant differences. Aston (28) produced confusing
evidence on this, stating that a reduction in glass panel size increases
the sound insulation at a given frequency; his results show the opposite effect. Pilon (29) measured transmission loss with a series of
equal-sized windows with different pane configurations. His results
are somewhat inconclusive, and may be explained on the doubtful
assumption that the lack of a horizontal axis of symmetry reduces the
window vibration.

Sound-absorbing layers adjacent to a panel. This is a subject which has
been given a fair amount of theoretical and practical treatment; it has

proved difficult to relate the two aspects and the varying experimental results due to the difficulty of measuring the internal properties of sound-absorbing material and probably to a lack of adequate control of the panel/absorbing layer interface in a known way. Lord (39) found that the addition of resonant sound absorbers within a transformer tank increased the amount of sound energy radiated from the tank. Beranek and Work (40, 41) reported on the sound insulation of a number of multilayer systems including sound absorbers. They concluded that the inclusion of sound absorption gives an improvement in sound insulation particularly at high frequencies where an undamped panel suffers from the effect of cavity resonances.

Northwood (42) has measured the sound insulation of a large number of plasterboard panels and many of his test panels included sound-absorbing layers. In general, the panels involving sound absorbers (rockwool or glasswool batts) show higher insulation than other similar panels without absorbers.

Mulholland (43) has shown that some sound-absorbing layers increase and some decrease the sound insulation. Rockwool showed consistent improvements in insulation greater than mass law expectations, particularly and significantly near the coincidence frequency of the base panel. On the other hand, expanded polystyrene was shown to reduce sound insulation usually over a range of frequencies about one octave wide. Flexible polyurethane gave results lying somewhere between these two extremes. The implication is that sound absorbers fall into groups; the first group includes such materials as rockwool, glasswool, and other fibrous materials and the second group is comprised of resonant absorbers, Helmholtz resonators, elastic layers, etc. Materials in the former group tend to increase sound insulation while the latter group can reduce the insulation near their resonant frequencies.

1.2.2. Double panels

Cavity width. It is difficult to measure the dependence of sound insulation on cavity width alone without introducing alterations in the mechanical linkage at the edges of the two panels. Some measured results show that there exists an optimum cavity width of about 100 mm which gives a maximum value for the transmission loss. Bruel (44) did not detect this effect in his early work but Cammerer and Durhammer (45) showed that the optimum did exist. Utley (46) showed that there is a maximum mean insulation at a cavity width of about 100 mm, followed by a minimum at about 200 mm. At larger cavity widths, the insulation increases.

Much work has been reported in the field of double glazing. Woolley (30) and Bazley (47) support Bruel (44) in showing that the insulation increases with cavity width. Brandt (48) has shown that the curve shape (transmission loss *vs* cavity width) depends on the glass panel thickness; a thicker glass pane causing the curve to level off at a smaller separation. Oosting (49) supports this view. Cook and Chrzanowski (12) report that in the limit as the panel separation exceeds the air wavelength of the incident sound, the insulation of a double wall tends to be twice the insulation of one of the single walls making up the double wall system. The conclusions are as follows. At small separations (such as one found with thermal double glazing) the mass–spring–mass resonance is about 200–400 Hz giving transmission losses approximately the same as a single panel of mass equal to that of one panel. As the separation is increased, the low frequency mass–spring–mass resonance falls below 100 Hz and becomes of less subjective importance; but the effects of standing wave resonances are seen at the higher frequencies. Ultimately when the separation exceeds the wavelength at a given frequency these resonances become unimportant and the system is left equivalent to two independent uncoupled leaves.

Absorbent cavity linings. There is some confusion in the reports of the effects of added absorption in the cavity. This may partly arise from an inadequate control of flanking transmission which is often a dominant factor, and from the fact that most wall constructions inherently have a finite amount of sound absorption in the cavity. The ostensible reason for the addition of absorbing materials in the cavity is to control the acoustic coupling between the leaves.

Cook and Chrzanowski (12) report that the addition of absorbing materials in heavy double walls with inherent flanking gives no improvement in insulation. When the flanking transmission was zero, an improvement of up to 15 dB at high frequencies was found.

The peripheral mounting of absorbents is discussed by Beranek and Work (40) and by Meyer (50).

Various materials such as sand, crushed brick, and tile chips have been placed in cavities; they mainly act as mass and friction elements.

London (51) reports that absorption in the cavity has little effect; Utley *et al.* (52), Ford *et al.* (53), and Constable (54) do not agree. The last three groups have also examined the effects of the distribution of sound absorption within cavities. Ford and Constable maintain that the maximum effect is achieved with the absorption located at velocity maxima of the mean modes within the cavity; Utley reports that within experimental error the increase in insulation is independent of the position of the absorption.

Ingemansson's (55) results show that lining the reveals of a window improves the insulation by some 3 dB at most frequencies; a not unexpected result when it is remembered that glass and most reveals have a low absorption coefficient. London (51), and Woolley (30) report little improvement with various forms of absorbent lining but there is evidence that the original cavity did not have very low absorption coefficient.

Eisenberg (27) reports an improvement of 2–4 dB when absorption is added to double glazing with a narrow cavity. Hochbrugge (22) introduced a layer of folded plastic between glass leaves and reports an improvement less than Eisenberg at high frequencies but greater at lower. Northwood (10) has studied a wide range of plasterboard panels and sound absorption in cavities and his results show a larger improvement in low-frequency sound insulation than has been reported by any other worker. Mulholland (56) has reported on the effect of completely filling cavities with sound-absorbing material; the effectiveness of sound-absorbing material in improving sound insulation is shown to be dependent on the mass of the original panel. The improvement is particularly marked with low-mass panels whereas heavy panels show no improvement at all on insertion of cavity absorption.

Panels with septa of different thicknesses. The motivation for the study of panels with elements of different thickness is the hope that the coincidence effects of the two elements will occur in different frequency bands. Ingemansson (55) shows that an improvement in insulation is found only near the critical frequencies and that it is only small. Brandt (48) reports that little is gained for glass panels under 4 mm thick for the critical frequency is beyond the measurement range, and Oosting (49) reports little effect. Eisenberg (27) compared 8+12+ 8 mm with 4+12+12 mm glass constructions; he reports that the transmission loss of the former is less than that of the latter by 4 dB for 45° incidence and 5 dB for 75° incidence. The Australian Building Research Centre (57) reports insulation is improved at medium and high frequencies by using constructions of different superficial masses.

Non-parallel panels. Based on simple models it has also been thought that by making separate leaves of the double wall non parallel, we might improve the insulation by distributing the effect of the mass–spring–mass resonance and the standing wave resonance. Oosting (49) shows no improvement over parallel walls but this might be expected in view of the diffuse nature of incident fields.

Mechanical linkage between panels. Although this is an important factor for our understanding, little published work is available. A demonstration of the effect of flanking and its control has been given by Beranek (58). Cook and Chrzanowski (12) report measurements on walls made with staggered stud construction and with normal stud construction. Aston (28) attempted to demonstrate the effect of mechanical linkage by wedging tiles between isolated frames; he reported a reduction of some 4 dB in the mean transmission loss. Brandt (48) reports that vibration isolation is no advantage in widths up to about 200 mm. Eisenberg (27) experimented with glass panes set in mortar and in rubber; a 1 dB difference was found in mean transmission loss and 3 dB near the coincidence region where one might expect the maximum effect.

1.3. THEORETICAL PREDICTIONS

The experimental work shows that the main factors determining the sound insulation of a panel are the mass per unit area of the panel and the frequency of the incident sound; the transmission loss of a panel increases by about 3 dB for each doubling of frequency and each doubling of the surface mass of the panel. Other properties of a panel and of the incident sound field also have their effect on sound insulation. The stiffness of a panel leads to coincidence phenomena which reduces the sound insulation in a range of frequencies. Multiple layers in a panel can introduce resonances of various types with a consequent reduction of insulation. However, the sound insulation of a complex panel can be increased above the mass law value by adding layers of sound-absorbing material, or vibration isolators, or air gaps, into the panel. However, it is difficult to increase the sound insulation of panels at low frequencies by such means. The sound insulation has been found to be not only a question of the performance of a panel alone but also a function of the performance of a system consisting of the sound source, the paths along which the sound has to travel, and the receiving environment.

1.3.1. The mass law

The basic mass law equation (59) is:

$$\tau = 1 / \left[1 + (\omega M \cos\theta / 2\rho c)^2 \right],$$

which is usually written:

$$TL = 10 \log \left[1 + \left(\frac{\omega M \cos\theta}{2\rho c} \right)^2 \right],$$

where ω = angular frequency, M = mass density of panel, ρ = density of air, c = speed of sound in air, τ = transmission coefficient, TL = transmission loss, and θ = angle of incidence.

It has been shown (58) that in a diffuse field, the first expression must be integrated giving the random incidence transmission factor:

$$\tau = \frac{\displaystyle\int_0^{\theta_L} \tau(\theta) \cos\theta \sin\theta d\theta}{\displaystyle\int_0^{\theta_L} \cos\theta \sin\theta d\theta}$$

This integral involves a parameter called the upper limiting angle (θ_L). The effect of θ_L on the sound insulation is to reduce the insulation below that of the normal incidence mass law ($\theta_L = 0$) which gives the greatest theoretical insulation value. For higher values of θ_L the insulation is reduced by a maximum of 10 dB when $\theta_L = 90°$. The nature of the sound field that can exist in a room determines the upper limiting angle. In a diffuse field, the sound power arriving at the panel as a function of angle of incidence will be independent of angle right up to grazing incidence ($\theta_L = 90°$). However, in any practical room a diffuse field will not exist. One of the differences between a diffuse field and non-diffuse one is that there is less energy arriving at the panel surface at a high angle of incidence in the non-diffuse field. This effect can be allowed for by the use of the upper limiting angle θ_L. In a laboratory facility $\theta_L = 80°$ is often found appropriate whereas in a domestic environment it is unlikely to exceed 70°. It is interesting to note that the nearer diffuse conditions are approached the lower the measured value of sound insulation.

1.3.2. The coincidence effect

While the mass law predicts with some success the approximate insulation of a wide variety of panels, many panels show a reduction of insulation below the mass law value over a range of frequencies. At frequencies above this minimum the sound insulation increases at a rate of about 12 dB/octave approaching the mass law value assymptotically. Cremer (60) explained the phenomenon in terms of dispersive forced bending waves travelling in the panel. He showed that when the velocity of the incident sound wave resolved along the panel equals the bending wave velocity in the panel, these bending waves build up to a high amplitude resulting in a large sound field on the far side of the panel. When the phenomenon occurs the insulation of a panel is reduced. Since the bending waves in the panel are dispersive the resolved velocities and wavelengths can only match above a certain frequency which is called the critical frequency. Further the matching is only effective over a few octaves since it takes place initially at grazing incidence at the critical frequency and at frequencies much higher the bending waves can only match air waves that are arriving almost normal to the panel. Cremer's theory was successful at predicting the critical frequency but not the detailed form of the sound insulation frequency curve. Recently this phenomenon has been investigated by Crocker (15) and by Sewell (61). These workers have described how coincidence takes place with finite size of panels and rooms and have concluded that at higher angles of incidence (where high room modes are coupled with high panel modes), the coupling coefficient is less than unity and an effective transparency of the panel is not expected. Theoretical sound insulation curves using these new ideas have shown better agreement with experimental curves.

1.3.3. Recent theoretical work

Advanced theoretical work on the subject of sound transmission of panels falls into two groups. Extensive work has been reported on the derivation of formulae for predicting the sound insulation of multi-layer panels. Of these, there are three principal methods; the work of Beranek and Work (40) extended to random incidence by Mulholland *et al.* (62) using the impedance transfer method; the approach of London (64) using a technique common in quantum physics and the multiple reflection or Fabry–Perot method and discussed by Mul-holland *et al.* (65), and Monna (66) and the modal theories and energy methods discussed by Lyon and Maidanik (67), Maidanik (68), and Crocker and Price (69).

Group 1: Infinite panel theories. The first of these two groups deals with layers which are assumed to be infinite in extent and of finite thickness. All three methods (62, 63, 65) of deriving a formula for multiple panels yield expressions which can be shown to be identical. This paper will consider the oblique incidence Beranek and Work (40) expression as derived by Mulholland, Price, and Parbrook (62). This method has the virtue of permitting the user to write down directly expressions for the insulation of a multiple panel consisting of any number of layers of a medium having a characteristic impedance ρc and an internal wave-number k.

The rules are discussed by Beranek and Work (40). Unfortunately the expression Beranek and Work derive cannot be compared with experimental values. Firstly finite septa do not behave in the way that the infinite panel theories assume and, therefore, any expression derived from infinite panel assumptions must be regarded as doubtful. This is further discussed in the next section on finite size panels. Secondly all the theories that appear to permit the prediction of the sound insulation of a multilayer panel are based on two assumptions:

1. That it is practically possible to measure or predict the specific impedance of any given material and also the complex wavenumber of sound waves transmitted through the material.

2. That it is possible to think of individual components of the panel each as a layer in which there is a constant acoustic impedance through the layer and discontinuities of acoustic impedance on the interface between adjacent layers and that this discontinuity produces reflected and transmitted waves. Away from these discontinuities, the transmitted wave is propagated with a given complex wavenumber.

The measurement or prediction of the properties of sound-insulating materials required by the above theories has proved difficult. Some attempts to predict properties of porous absorbents from basic geometrical and acoustical properties of the materials and also attempts to measure these properties using standing-wave tube techniques are reported by Beranek (70), Zwikker and Kosten (71), Scott (72), and Pyett (73), including oblique incidence measurements using a rectangular wave guide. Of these workers, Scott (72) checked the consistency of the measurements he used in predicting the sound absorption of thicker layers of a similar material and reported good agreement.

The inherent accuracy of these current methods has so far been insufficient to give values of impedance that are reproducible to any great extent. Current research on this topic seeks more and more accurate methods of measurements of these complex coefficients; this has been reported on by Bokor (74), Walker (75), and Flockton (76). On the second assumption it has been shown by Attenborough (77) that fibrous materials do not behave as if they were layers of fixed specific impedance and wavenumber except at normal incidence. As sound passes through such layers it undergoes multiple scattering by the individual fibres and the macroscopic effect of this is to produce exponential decay of the incident wave with corresponding growth of forward and backward scattered waves arising within the material. This approach cannot be assimilated into current multilayer theories which will need extensive modification in the light of the new ideas. Work in this direction has been reported by Mok (78).

Group 2: Modal theories and energy methods. In the field of aerospace technology the problem of the response of structures to pressure fluctuation (or incident sound fields) has received much attention. Powell (79), Skudrzyk (80), Lyon, Dyer, and Ribner (81–4) were the first to formulate theories for the response of structures to random pressures. Of these theories that of Powell seems to have become the best known. This work mainly concentrated on the subject of aircraft but it was extended in the direction of building structures by such workers as Lyon (85) and Maidanik (68). Then the possibilities became clearer and the new technique was brought directly to bear on the problem of sound insulation by White and Powell (86) and Crocker and Price (69).

Already the difficulties in the infinite panel theories were becoming apparent. The failure of Cremer's theory to predict exactly the fine structure of coincidence curves was known and also it had been shown (87) that the vibration amplitude of simple panels was much greater than would be expected from mass law considerations. Using the new techniques, Crocker (88) was able to show that vibration amplitude of simple panels was expected to be greater than that of mass law panels. In fact, the theory was able to predict the exact form of the deviation of vibration amplitude from its theoretical mass law value over the whole range of frequencies including the coincidence range.

A new conception of coincidence has resulted from this work (89). The new ideas add to the work of Cremer by extending coincidence theory to finite-size plates coupled to finite rooms and lead to a better fit with observed coincidence curves than has hitherto been possible. The work is still in active development and has stimulated new lines of experimental work such as that at present being carried out by Yaneske and Walker (90) on active damping of panel modes by means of transducers and electronic feedback devices. The difficulty of damping all modes as they run through their frequency/phase relationship is reported to be more apparent than real and Yaneske and Walker hope that considerable increases in sound insulation may be obtained over a useful frequency range.

1.3.4. Vibration flanking and transmission through ties

Recent theoretical developments of the work of Cremer (91) on the transmission of a vibration through connected structures started by Lyon *et al.* (67) and Eichler (92), show potential in considering the problems with vibration flanking and the transmission through ties connecting panel structures. This work has been successfully applied in the case of dynamic structures and it should be possible to extend this to building constructions but little work has been reported.

Kihlman (93) also starting from the ideas of Cremer has produced some detailed theoretical work on the transmission of structure-borne sound around corners and between adjacent rooms. Kihlman quotes theoretical values of vibration insulation for specific cases and obtains fairly good agreement with experiments. This problem is also being studied by Zaborov (94), Crocker (95), and Bhattacharya (96).

ACKNOWLEDGEMENTS

We wish to thank Mrs J. A. Marsh of Pilkington Brothers Ltd Environmental Advisory Service, and Dr K. Attenborough, Department of Building Science, University of Liverpool, for helpful advice and comment.

Appendix

MEASUREMENT OF SOUND INSULATION

The standard method of measuring airborne sound insulation of a construction is given in ISO R140: 1960 (98). This is based largely on similar codes used in several countries. ISO R140 is mainly concerned with laboratory testing; for field tests the recommended procedures are similar to the laboratory tests although it is probable that further recommendations will be made for field tests. Since there is an immediate and pressing need for field tests, a temporary standardized method is argued to be better than none.

The sound transmitted between two rooms depends not only on the insulation properties of the party wall but also primarily on the flanking transmission, the area of the wall, the amount of absorption in the receiving room, and the nature of the sound fields in the two rooms. The aim of ISO R140 is to set out a procedure to determine the insulation of the party wall alone. This is done in the laboratory by inserting a test wall in an opening between two vibration isolated rooms in which one aims to have reverberant diffuse fields. ISO R140 recommends test conditions concerning the volume and shape of the reverberant rooms, the exclusion of flanking sound, the size and mounting of the panels, the generation and measurement of the sound field, and the frequency range.

If L_1 is the average sound pressure level in the source room and L_2 that in the receiving room, then the transmission loss or sound reduction index is:

$$TL = L_1 - L_2 + 10 \log_{10}(S/A),$$

where S is the area of the panel under test and A the total absorption area in the receiving room. A can be found by measuring the reverberation decay rate as indicated in ISO R354 (99).

Errors can occur where TL is less than 15 dB from a coupling of the two fields in the source and receiving rooms through the common wall. These may be avoided by using a modified technique (100).

The recommended method is known to be imperfect. Measurement made by the same person on a panel at different times and by different people on the same panel show considerable variations, more so at the lower than at the higher frequencies: 8 dB differences at the lower frequencies and 2 dB at the higher are not uncommon. Such unexplained deviations are embarrassing in experimental work. However, the position is much more difficult when we are measuring to see if a construction meets a grading curve. For questions of economics and weight often mean that the construction is unlikely to more than just meet the requirement and deviations of this order could fail a construction.

The main difficulty lies in meeting the requirement for isotropic and diffuse sound fields in both source and receiving rooms. It is doubtful if we could ever achieve such an ideal field and one has to resolve what deviations from this requirement can be accepted. Some work is in progress on these matters and has been reported in various papers including the temperature effect on the modal response of rooms. [See, for example, the BRS Symposium, December 1967, *Sound insulation measurements and the Building Regulations* (101).]

Field measurements are more difficult than those in laboratory and we meet the problems mentioned above—often considerably aggravated. In addition there are problems of inadequate room size and shape, over-damped reverberant conditions, flanking transmissions, and background noise. For grading purposes, one often only has a single party wall between two semi-detached houses on which four measurements have to be made and the results examined on a pass/ fail basis against a grading curve specified explicitly or implicitly in the Building Regulations.

For comparison purposes, insulation measurements are often normalized either for a reverberation time of 0·5 s or 10 m² absorption in the receiving room.

It is likely that it will be some while before our difficulties and problems are resolved. Perhaps one can refer the reader to one new approach to field measurements (97) as an indication of the type of rethinking that may be necessary.

REFERENCES

1. BERANEK, L. L. (1957). 'Revised criteria for noise in building', *Noise Control*, **3**, 19–27.
2. INTERNATIONAL STANDARDS ORGANISATION: Draft 314, 'Noise rating with respect to annoyance, speech communication and hearing conservation'.
3. BRITISH STANDARDS INSTITUTION. BS 4142: 1956.
4. *The Building Regulations, 1965, Part G.* (HMSO).
5. *The Building Standards (Scotland) Regulation, 1963*, no. 1897 (S102) (HMSO).
6. INTERNATIONAL STANDARDS ORGANISATION. ISO R717: 1968.
7. EISENBERG, A. (1958). *Glastechnische Berichte*, **31**, 297–302.
8. UTLEY, W. A., and FLETCHER, B. L. (1969). 'Influence of edge conditions on the sound insulation of windows', *Appl. Acoust.*, **2**, 131–6.
9. PARKIN, P. H., PURKIS, H. J., and SCHOLES, W. E. (1960). *Field measurements of sound insulation between dwellings* (London: HMSO).
10. NORTHWOOD, T. D. (1968). 'Transmission loss of plasterboard walls', *Building Research Note 66*, Nat. Res. Council of Canada (Ottawa).
11. TNO and TH Delft Report No. 58.636 (1958).
12. COOK, R. K., and CHRZANOWSKI, P. (1957). *Handbook of noise control* (ed. Harris), chap. 20 (New York: McGraw-Hill).
13. 'Airborne sound insulation of glass' (in press) (St Helens, Lancs: Pilkington Bros. Ltd).
14. CREMER, L. (1949). 'Sound insulation of panels at oblique incidence', *Physical Society, Acoustics Group Symposium*, p.23.
15. CROCKER, M. J. (1969). *Response of structures to acoustic excitation*, PhD thesis, University of Liverpool.
16. *Comparison between a laminated glass and glass of the same thickness.* Pilkington Bros Ltd. Internal works report (April 1959).
17. KURTZE, G., TAMM, K., and VOGEL, S. (1955), *Acustica*, **5**, 223.
18. WESTPHAL, W. *Acustica*, **4** (1954), 603–10 and 1 (1957), 335–48.
19. KURTZE, G. (1958). *Q. Progr. Report*, April/June, BBN Report to Office of Naval Research, USA.
20. SHARP, B., and BEAUCHAMP, J. W. *BRS Note*, EN.119/67 (Watford).

21. UTLEY, W. A. (1967). *The transmission of sound through double and triple panels*, PhD thesis, University of Liverpool.
22. HOCHBRUGGE, G. (1963). *Kamf dem Larm*, **10** (6), 152–5.
23. GOMPERTS, M. C., and KIHLMAN, T. (1967). *Acustica*, **18**, 144–50.
24. INGERSLEV, F., and NEILSEN, A. K. (1944). *Acoustical Laboratory of the Danish Academy of Technical Sciences Publ. No. 1.*
25. WILSON, C. P., and SOROKA, W. W. (1965). 'Approximation to the diffraction of sound by a circular aperture in a rigid wall of finite thickness', *Jl Acoust. Soc. Am.* **37**, 286.
26. MULHOLLAND, K. A., and PARBROOK, H. D. (1967). 'Transmission of sound through apertures of negligible thickness', *Jl Sound Vib.* **5**, 499.
27. EISENBERG, A. (1961). *Glastechnische Berichte*, **34**, 544–74.
28. ASTON, G. H. (1949). *Report of 1948 Symposium of Acoustics Group, Noise and Sound Transmission 7–15* (London: The Physical Society).
29. PILON, J. M. (1967). *Materiaux de Construction*, p. 95 (Paris: St Gobain).
30. WOOLLEY, R. M. (1967). *Glass Age*, **10** (May) 44.
31. OBERST, H. *Acustica* (*Akust. Beih 4*), **2** (1952) 181; *Acustica*, **6** (1956), 144.
32. ——, and BECKER, G. W. (1954). Ibid. **4**, 433.
33. LIENARD, P. *Recherche Aeronaut*, **20** (1951), 11; *Ann Telecom.* **12** (1957), 359.
34. KURTZE, G., and WATTERS, B. L. (1959). 'New wall design for high transmission loss or high damping', *Jl Acoust. Soc. Am.* **31**, 739.
35. KERWIN, E. M. (1958). BBN Report 547 to Convair.
36. MOSS, D., KERWIN, E. M., and DYER, I. (1958). BBN Report 565 to Office of Naval Research, USA.
37. LIBBY-OWENS-FORD. *Sound transmission through glazing materials and glass and the reduction of airborne sound.* Bulletins by Arver, W. J., and Gwyn, J. D.
38. CUMMINGS, A. (1958). *The mechanism of sound transmission through single and double panels*, PhD thesis, University of Liverpool.
39. LORD, P., University of Salford. Private communication.
40. BERANEK, L. L., and WORK, G. A. (1949). 'Sound transmission through multiple structures', *Jl Acoust. Soc. Am.* **21**, 419–28.
41. BERANEK, L. L. (1960). *Noise reduction*, p. 382 (New York: McGraw-Hill).
42. NORTHWOOD, T. D. (1968). 'Transmission loss of plasterboard walls', *Building Research Note 66*, Nat. Res. Council of Canada (Ottawa).
43. MULHOLLAND, K. A. (1969). 'The effect of sound absorbing materials on the sound insulation of single walls', *Appl. Acoust.* **2**, 1.
44. BRUEL, P. V. (1951). *Sound insulation and room acoustics*, chap. 5 (London: Chapman and Hall).
45. CAMMERER and DURHAMMER. (1934). 'Propagation of airborne sounds through solid walls with air spaces', *Gesundheits Ingenieur*, **57**, 556.
46. UTLEY, W. A., and MULHOLLAND, K. A. (1968). 'The transmission loss of double and triple walls', *Appl. Acoust.* **1**, 15–20.
47. BAZLEY, E. N. (1966). *The airborne sound insulation of partitions* (National Physical Laboratory, HMSO).
48. BRANDT, O. (1954). *Tehnisk Tidskrift*, 1129–33 (Dec.).
49. OOSTING, W. A. *et al.* (1967). *Investigation of sound insulation of flat glass*. TNO and TH Delft Report No. 706-007 (April).
50. MEYER, E. E. (1935). *Elek Nachr Tech.*, p. 393.
51. LONDON, A. (1950). *Jl Acoust. Soc. Amer.* **22**, 270–9.
52. UTLEY, W. A., CUMMINGS, A., and PARBROOK, H. D. (1969). *Jl Sound Vib.* **9**, 90–6.

53. FORD, R. D., LORD, P., and WILLIAMS, P. C. (1967). Ibid. **5**, 22–8.
54. CONSTABLE, J. E. R. (1936). *Proc. Phys. Soc.* **48**, 690–8.
55. INGEMANSSEN, S. (1967). *Byggmasteren*, **2**, 58–64.
56. MULHOLLAND, K. A. 'Sound insulation measurements on a series of double plasterboard panels with various infills', *Appl. Acoust.* (to be published).
57. Australian Window Glass: Commonwealth Building Station, sponsored investigation 42 BS.50/97, 1966.
58. BERANEK, L. L. (1960). *Noise reduction*, chap. 13 (New York: McGraw-Hill).
59. —— Ibid., p. 296.
60. CREMER, L. (1942). 'Theorie der Schalldammung dunner Wande bei Schragem Einfall', *Akust. Zeit.* **7**, 81.
61. SEWELL, E. C. Building Research Station, England. Internal notes: In26/65 138/65.
62. MULHOLLAND, K. A., Price, A. J., and PARBROOK, H. D. (1968). 'Transmission loss of multiple panels in a random incidence field', *Jl Acoust. Soc. Am.* **43**, 1432.
63. LONDON, A. (1949). 'Transmission of reverberant sound through single walls', ibid. **42**, 605.
64. —— (1950). 'Transmission of reverberant sound through double walls', ibid. **20**, 270.
65. MULHOLLAND, K. A., PARBROOK, H. D., and CUMMINGS, A. (1967). 'The transmission loss of double panels', *Jl Sound Vib.* **6**, (3), 324.
66. MONNA, A. F. (1938). 'Absorption of sound by porous walls', *Physica*, **5**, 129.
67. LYON, R. H., and MAIDANIK, G. (1962). 'Power flow between linearly coupled oscillators', *Jl Acoust. Soc. Am.* **34** (5), 623.
68. MAIDANIK, G. (1962). 'Response of ribbed panels to reverberant acoustic fields', ibid. **34** (6).
69. CROCKER, M. J., and PRICE, A. J. (1969). 'Sound transmission using statistical energy analysis', *Jl Sound Vib.* **9** (3), 469.
70. BERANEK, L. L. (1947). *Jl Acoust. Soc. Am.* **19**, 556.
71. ZWIKKER, C., and KOSTEN, C. W. (1949). *Sound absorbing materials* (Elsevier).
72. SCOTT, R. A. (1946). *Proc. R. Soc.* **58**, 358.
73. PYETT, E. T. (1953). *Acustica*, **3**, 375.
74. BOKOR, A. (1969). 'Attenuation of sound in lined ducts', *Jl Sound Vib.* **10**, 10.
75. WALKER, C. British Gypsum Ltd, East Leake Notts (private communication).
76. FLOCKTON, S. F., and MULHOLLAND, K. A. (1970). 'Investigation of discrepancy between predicted and measured values of the sound insulation of layered panels', *Jl Sound Vib.* **11**, 275–80.
77. ATTENBOROUGH, K. (1969). *Sound dissipation in porous media*, PhD thesis, University of Leeds.
78. MOK, C-H. (1969).)'Effective dynamic properties of a fibre-reinforced material and the propagation of sinusoidal waves', *Jl Acoust.. Soc. Am.* **46**, 631.
79. POWELL, A. (1958). 'Comments on the response of a string to random distributed forces', ibid. **30**, 365.
80. SKUDRZYK, E. (1958). 'Vibration of a system with a finite or an infinite number of resonances', ibid. **30** (12), 1140.
81. LYON, R. H. (1956). 'Propagation of correlation functions in continuous media', ibid. **28**, 76.
82. —— (1956). 'Response of strings to random noise fields', ibid. **28**, 391.
83. DYER, I. (1959). 'Response of plates to decaying pressure field', ibid. **31**, 922.
84. RIBNER, H. S. (1956). 'Boundary-layer-induced noise in the interior of aircraft', *UTIA Report*, No. 37.

85. LYON, R. H. (1963). 'Noise reduction of rectangular enclosures with one flexible wall', *Jl Acoust. Soc. Am.* **35** (11), 1791.
86. WHITE, P. H., and POWELL, A. (1965). 'Transmission of random sound and vibration through a rectangular double wall', ibid. **40,** 821.
87. UTLEY, W. A., and MULHOLLAND, K. A. (1967). 'Measurement of transmission loss using vibration transducers', *Jl Sound Vib.* **6,** 419.
88. CROCKER, M. J. (1969). *The response of structures to acoustic excitation and the transmission of sound and vibration*, fig. 63, PhD thesis, University of Liverpool.
89. BHATTACHARYA, M. C. 'Coincidence effect in finite panels' (Internal memo BS/A/70/2), University of Liverpool.
90. YANESKE, P. P., and WALKER, L. A., University of Leeds (private communication).
91. CREMER, L. (1953). 'Calculation of sound propagation in structures', *Acustica*, **3** (5), 317.
92. EICHLER, E. (1965). *Jl Acoust. Soc. Am.* **37** (6), 995.
93. KIHLMAN, T. (1967). 'Sound radiation into a rectangular room, applications to airborne sound transmission in buildings', *Acustica*, **18,** 1.
94. ZABOROV, V. I. Scientific Research and Planning Institute of Construction Materials, Chelyabinsk, USSR.
95. CROCKER, M. J. Department of Mechanical Engineering, W. Ray Herrick Laboratories, Purdue University, USA.
96. BHATTACHARYA, M. C. Department of Building Science, University of Liverpool.
97. SCHULTZ (1965). *Proc. 5th International Congress on Acoustics*, (Paper F34, Liege).
98. ISO R140: *Field and laboratory measurements of airborne and impact sound transmission.*
99. ISO R354: *Measurement of absorption coefficient in a reverberation room.*
100. MULHOLLAND, K. A., and PARBROOK, H. D. (1965). 'The measurement of sound transmission loss of panels with small transmission loss', *J. Sound Vib.* **2,** 502–9.
101. BRS Symposium (Dec. 1967). *Sound insulation measurements and the Building Regulations* (Garston, Watford, Herts: Building Research Station).

2

Plastics and plastics-based composites

Albert G. H. Dietz
Professor of Building Engineering,
Massachusetts Institute of Technology

2.1. INTRODUCTION

Plastics and composites based upon or containing plastics as essential ingredients are dealt with in this chapter, with especial emphasis being placed on their use in buildings.

A precise definition of plastics is not easy to formulate. They are, indeed, plastic at some stage in their histories; that is, they are soft, malleable, or even liquid, and can be shaped, sometimes by casting, usually by pressure, heat, or both. Some, called thermoplastic, can be softened any number of times by re-heating; others, called thermo-setting, are soft only once, but when hardened, remain hard, and heating may only make them harder. Being plastic, however, is not enough; many materials not considered to be plastics are malleable at some stage. Plastics are restricted to organic materials, based largely upon carbon chemistry, but many naturally occurring materials, such as asphalt, are organic, but not considered to be plastics. Plastics are, therefore, restricted to synthetic organic materials. Finally, they are high polymers, that is, very large molecules in a chain-like or network configuration made up of many small repeating molecular aggregations or monomers that have been connected to form the high polymer.

There are some arbitary exclusions and some borderline cases. Synthetic rubber, for example, has all of the characteristics mentioned, but is not considered to be a plastic. Silicones, on the other hand, have a silicon–oxygen backbone, and carbon is found in the side groupings, but silicones are included in the family of plastics.

Although many polymeric materials have been synthesized in the laboratories, there are perhaps 20–30 major classes of commercially important plastics. But modifications of many kinds brought about by the use of fillers and plasticizers, and crossing or copolymerizing several classes into the same polymer, provide a great range of properties and resistance to different environments.

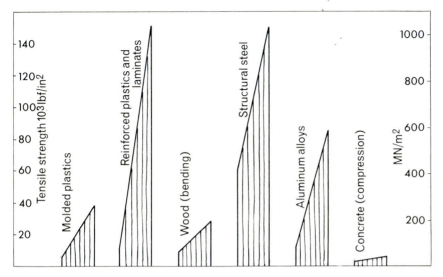

Fig. 2.1. Tensile strength of plastics compared with other materials

2.2. PROPERTIES

Among the most important properties for engineering and architectural applications are strength, stiffness, toughness, hardness, optical characteristics, thermal properties, durability, and resistance to fire (Table 2.1).

2.2.1. Strength

Strength of pure unmodified plastics varies over a wide range, from weak for some soft materials, to considerably higher than wood and concrete for others. When fillers are added, the strength is increased considerably, and when plastics are combined with tough constituents such as strong fibres in composites (see below), the materials attain a very high strength and, on a strength-to-weight basis, are among the foremost engineering materials available. Ranges of tensile strength of moulded and reinforced plastics, including laminates, are shown and compared with several other materials in Fig. 2.1. Compressive and flexural strengths are generally higher.

Table 2.1. *Properties of plastics*

	Compressive strength 10^3 lbf/in² (MN/m²)	Flexural modulus 10^3 lbf/in² (MN/m²)	Compressive modulus 10^3 lbf/in² (MN/m²)
Acrylic	11–19 (75·7–131)	390–470 (2690–3240)	390–460 (2690–3170)
ABS	4·5–7 (31·2–48·2)	—	—
Phenolics	12–15 (82·8–104)	—	—
Polycarbonate	12·5 (86·2)	340 (2350)	34·5 (238)
Polyethylene (low density)	—	8–60 (55·2–414)	—
Polyethylene (high density)	3·2 (22·1)	100–260 (689–1790)	—
Polypropylene	5·5–8 (37·9–55·2)	170–210 (1170–1450)	150–230 (1030–1590)
Polystyrene	4–9 (27·6–62)	—	—
Polyvinyl chloride (rigid)	8–13 (55·2–89·7)	—	—
Polyurethane	20 (138)	—	—
Nylon (type 6/6)	6·7–12·5 (46·2–86·2)	395–410 (2720–2830)	245–8 (1690–1710)

Architectural and Engineering News, from data supplied by the Society of the Plastics Industry.

2.2.2. Stiffness

Stiffness, or modulus of elasticity, of most plastics materials is low in comparison with other materials such as steel, aluminium and glass, and is more nearly in the range of wood and concrete, being well below these in many instances (Fig. 2.2). When modified by fillers, especially fibrous fillers, or when combined with high-strength, high-modulus fibres and sheets, the stiffness or elastic modulus may be

Resistance to heat °F (°C)	Flammability	Effect of sunlight	Clarity
140–200 (60–93)	Slow	None	Transparent (>92 per cent light transmission), translucent and opaque (haze <3)
170–210 (77–99)	Slow	None to yellows slightly	Translucent to opaque
160 (71)	Very slow	Colours may fade	Transparent, translucent, opaque
250 (121)	Self-extinguishing	Slight colour change	Transparent to opaque
180–212 (82–100)	Very slow	Requires black for complete protection	Transparent to opaque
250 (121)	Very slow	Requires black for complete protection	Transparent to opaque
250–320 (121–60)	Slow	Requires black for complete protection	Transparent, translucent, opaque
140–75 (60–80)	Slow	Some strength loss	Translucent to opaque
150–75 (66–80)	Self-extinguishing	—	Transparent to opaque
190–250 (88–121)	Slow	None to yellow	Clear to opaque
175–250 (80–121)	Self-extinguishing	Discolours slightly	Translucent to opaque

markedly increased, but even the best of such composites are still below aluminium and glass, for example, and well below steel. As is brought out later, stiffness is best achieved by employing inherently stiff shapes.

Both strength and stiffness of some classes of plastics, notably the thermoplastics, are temperature and time dependent. As the temperature rises, the materials become softer and weaker, and as it drops, strength and stiffness increase. Similarly, as is true of many other

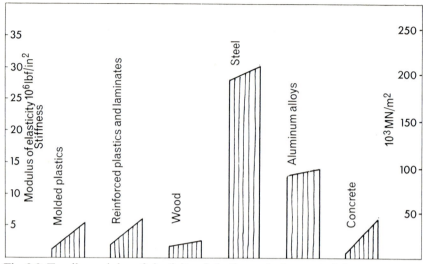

Fig. 2.2. Tensile modulus of elasticity of plastics and other materials

materials, under long-continued loads, plastics may continue to deflect, or creep. Whether the changes in strength and stiffness, or creep under continued loads, are important, depends upon individual circumstances. Commonly, they can be taken into account in engineering design. These characteristics are less marked among thermosets than among thermoplastics.

2.2.3. Toughness

Toughness is difficult to define and measure. Generally, it is taken as a measure of the ability of a material to absorb mechanical energy, and is likely to be measured by empirical impact devices such as a swinging pendulum or a falling ball. A great range of values is found among plastics; some absorb a great deal of energy and others do not, i.e., they are relatively brittle. Some brittle plastics can be made tough by incorporating modifiers such as plasticizers and fillers, or by copolymerization. Some plastics, considered relatively brittle as plastics, are tough compared with other materials. Under a standard falling steel ball test, for example, $\frac{1}{8}$ in (3·2 mm) thick clear acrylic is some 25–30 times as impact-resistant as standard window glass of the same thickness.

2.2.4. Hardness

Hardness, like toughness, is generally measured by empirical means that may be difficult to interpret in general terms. When scratched by a standard sharp point under a standard load, plastics are not so hard as glass or most metals, but materials such as the melamine formaldehyde finish on high-pressure laminates are harder than commonly employed lacquers and varnishes. Similarly, when plastics are indented by standard spherical or conical indentors under standard loads, they are softer than typical ceramics such as glass and burnt clay, but may be harder than most species of wood across the grain, again depending upon type of plastic and formulation.

2.2.5. Optical characteristics

Plastics may range from high clarity to complete opacity. The acrylics, for example, are among the clearest materials available, with light transmission in the visible range approaching that theoretically possible on the basis of index of refraction. Transmission in the near ultra-violet may be high or low, depending upon formulation. Transmission in the infra-red is variable and decreases as wavelength increases, with a cut-off at approximately 2·5–3·0 μm. These clear plastics, therefore, behave much the same as equally clear glass. For example, they exhibit the same 'greenhouse' effect, that is, they are transparent to incoming solar radiation into a space, but opaque to long-wave outward radiation from objects heated by the solar radiation within the space (Fig. 2.3).

Some plastics materials are little affected by sunlight and maintain their clarity for many years; others are degraded, darken with time, and may craze or develop a network of fine cracks, but may be satisfactory for artificial light. Care must, therefore, be exercised in selecting plastics for lighting.

The clear plastics can be coloured by the addition of dyes for transparent colours, or pigments for varying degrees of colour and opacity. Other plastics have distinctive colours, and their further colourability is, therefore, limited.

The ready formability of acrylic and other clear plastic rods makes it possible to 'bend' light along such rods, provided the radius of curvature of the bends is not so small as to exceed the internal critical reflecting angle limited by the index of refraction of the plastic.

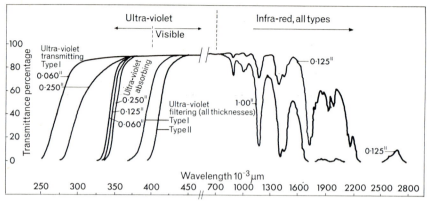

Fig. 2.3. Transmittance of acrylic plastic as function of wavelength

Internal carved figures can be made to glow by edge lighting transmitted by the plastic but scattered by the carving.

2.2.6. Thermal characteristics

Coefficient of expansion. Unmodified plastics characteristically have high coefficients of thermal expansion, a feature that must be allowed for in design of plastics parts subject to marked changes in temperature. Usually, thermal expansion is several times as high as that of aluminium, for example. Fillers may reduce this markedly. Strong fibres such as glass have lower thermal coefficients, and when incorporated with plastics in composites, reduce the over-all coefficient to values more nearly like those of aluminium alloys. This is helpful when such materials are employed in large parts such as components of buildings (Fig. 2.4).

Heat transmission. Compared with metals, or even with concrete and glass, solid plastics are moderately good heat insulators, with relatively low heat transmission coefficients (Fig. 2.5a). Most plastics can be foamed and many are. In this form, they are among the best insulators available for the temperature ranges in which they may be employed (Fig. 2.5b). Coefficients depend upon the density of the foam and the cell size. For many applications, foams such as polystyrene and polyurethane are used with densities as low as 1 lb/ft^3 (16 kg/m^3) to 2 lb/ft^3 (32 kg/m^3), or about 1–3 per cent of the solid

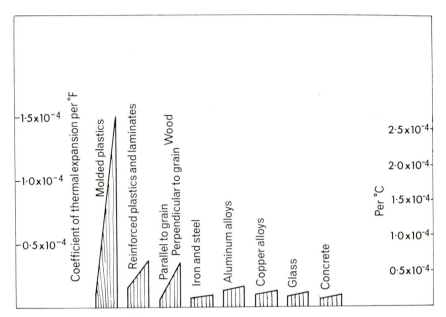

Fig. 2.4. Coefficient of thermal expansion of plastics and other materials

density. In this range, coefficients of conductivity, are approximately 0·20 Btu in/ft²h °F (0·29 w/m °C). When polyurethane is blown with a fluorocarbon gas, the coefficient may be approximately one-half that value.

If foams are made with large cells, appreciable convection currents may be set up within them by temperature differences, and the insulating value may be reduced. There is, therefore, a lower density limit for foams, below which coefficients of heat transfer tend to rise. This is lower than the density ranges noted above.

When foams need to have strength, stiffness, hardness, or resistance to abrasion, the density may be increased, with some loss in insulating value.

2.2.7. Moisture permeability

Many plastics are employed as films for packaging or other protective purposes. Permeability to water, water vapour, or other gases is, therefore, important. This property varies considerably with different

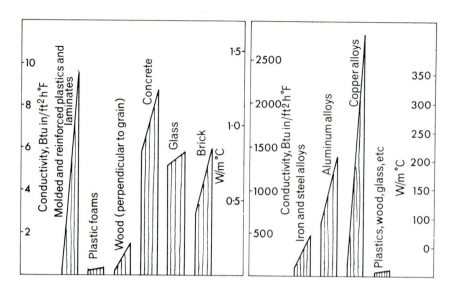

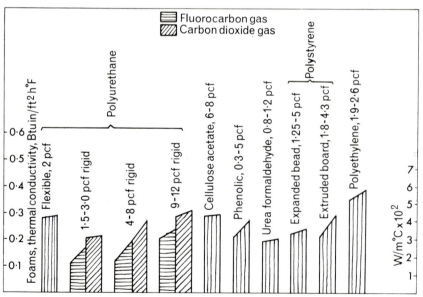

Fig. 2.5. (*a*) Thermal conductivity of plastics and other materials. (*b*) Thermal conductivity of plastics foams

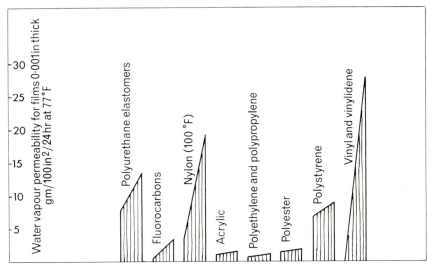

Fig. 2.6. Water vapour permeability of selected plastics

materials and formulations (Fig. 2.6). Polyvinyledine chloride, acrylics, fluorocarbons, polyethylene, and polypropylene have low permeability; whereas certain polyvinyl chloride and nylon formulations are relatively high. Depending upon the application, therefore, different films may be chosen for different purposes.

2.2.8. Durability

Depending upon the exposure conditions and the formulation of a given material, plastics may be highly durable or may deteriorate rapidly. The great majority of plastics are unaffected by water and may, therefore, be employed as containers without fear of corrosion. Fluoroplastics are notably inert to solvents, whether inorganic or organic, aliphatic or aromatic. Other plastics are inert to many, but not all, solvents. They are, therefore, commonly employed for handling corrosive chemicals of many kinds.

For exposure to outdoor conditions, some plastics have established a good record of durability, whereas others have not. Because of their relative newness, many have not yet had a long history of exposure and none can compare in age with wood, glass, and stone, or even aluminium. A few, such as acrylics and polyvinyl chloride, have been employed on the exteriors of buildings for periods of twenty-five years or more and have given good results. Fluorocarbons have shown excellent resistance to sunlight under extreme conditions. Silicones have stood up well. Polycarbonates and cellulose acetate butyrate are employed outdoors. Other plastics depend upon formulation. Unmodified polyethylene, for example, deteriorates rapidly when exposed to sunlight, but when carbon black is added, its resistance to outdoor exposure is excellent.

Assessment of durability for many conditions, such as weathering, is hampered by the lack of reliable short-time accelerated tests that will accurately predict long-time performance in the field, unless those tests are accompanied by a history of long-time exposure. For many plastics such a history is not available and considerable uncertainty, therefore, surrounds their expected behaviour under long-time exposure.

2.2.9. Fire-resistance

Plastics are organic materials and can, therefore, be destroyed by hot enough fires. Some do not ignite; some are self-extinguishing; others burn slowly or rapidly. Basic composition and the presence of modifiers such as fillers and plasticizers have a marked effect upon flammability. The chemical constituents of many plastics are similar to those of wood and fabrics, and the combustion products are also similar. With plentiful oxygen, these are mainly carbon dioxide and water; with deficient oxygen, carbon monoxide and smoke are likely to be given off. If other elements such as chlorine and nitrogen are present in the plastic, they will occur in the combustion products.

Burning rate and fire-resistance are commonly measured by a number of empirical tests. The simplest is to place a Bunsen burner under one end of a small horizontal bar of the material and see if it ignites; if it does, the rate at which it burns is noted. A test for rate of burning of surface materials is a horizontal tunnel in which the tested material forms the under-surface of the roof. Rate of propagation of flame along the tunnel, and amount of smoke evolved, are measured and compared empirically with red oak. Considerable research is being conducted to find other tests that may be considered more satisfactory and realistic than the standard tests commonly employed.

2.2.10. Examples of non-structural uses of plastics in building

Uses of plastics in non-structural applications in buildings have become so numerous, albeit modest in total volume, that it is impossible to list them all. A few examples will illustrate.

Illumination. Old-style skylights, consisting of sheets of glass puttied into frames, are hard to keep watertight and have been largely supplanted by bubble-shaped acrylic one-piece domes. These are easily fabricated by heating the sheets to the softening point, clamping them along their edges over a suitably shaped opening in a vacuum tank, and drawing a vacuum, thereby draw-stretching the sheet into the opening into the shape of a dome. The sheet cools and hardens in this shape.

Thermo-formed sheets, transparent or translucent, are employed widely in artificial illumination and for lighted signs. They are especially welcome for street lights where vandalism causes frequent breakage of glass. Acrylics and polycarbonates are among the materials favoured because of their toughness. Flat sheets are employed for glazing where breakage hazard is high, even though they scratch more easily, and allowance must be made for their high coefficients of thermal expansion.

Flooring. Resilient flooring sheets and tiles are dominated by plastics, particularly polyvinyl chloride and the coumarone-indene resins (Fig. 2.7). Colours and textures are virtually unlimited, and many fillers, such as asbestos, are commonly included to modify properties and reduce costs. Because they are thermoplastic, they may be heated and bent into corners, for example. By the same token, if warmed, they may be indented by heavy furniture or spike heels.

Wall covering. Flexible sheets of polyvinyl chloride, sometimes backed with fabric, paper, or flock, are applied to interior wall and partition surfaces to supply a decorative and wear-resistant surface. High-pressure decorative laminates (see below) are also employed for wall covering. Fig. 2.8 illustrates the use of polyvinyl chloride on exterior walls.

Fig. 2.7. Resilient polyvinyl chloride flooring
Armstrong Cork Company

Fig. 2.8. Exterior siding of polyvinyl chloride
Bird & Son

Fig. 2.9. Prefoamed polystyrene foam boards insulation under concrete slab
Dow Chemical Company

Fig. 2.10. Polyethylene vapour barrier under concrete slab

Union Carbide Corporation

Fig. 2.11. Polyvinyl chloride plumbing lines

B. F. Goodrich Chemical Co.

Insulation. As Fig. 2.5*b* shows, foamed plastics exhibit low thermal conductivity and are, therefore, used as thermal insulators in buildings. They may be (1) prefoamed and cut into slabs and boards; (2) made into beads containing a latent volatile substance which changes to a gas at moderately elevated temperatures and causes the beads to expand and become fluffy; or (3) mixed as a liquid whose constituents react to give off a gas and cause the mass to rise while it stiffens into a resilient or hard foam. The first two are exemplified by polystyrene, the last by polyurethane. Boards and slabs are applied in the field, or may have sandwich facings bonded to them; the expandable beads and foaming liquid are foamed in place. They are especially suitable for irregular spaces otherwise hard to fill (Fig. 2.9).

Vapour barriers. If water vapour passes from a warm zone to a cold zone in a wall or roof, for example, it may become chilled, reach the dew point, and condense, with consequent deterioration of the building component. To prevent this, vapour barriers may be interposed to stop the flow of vapour to the condensation zone. Various materials such as metal foil and some plastics films are commonly employed (Fig. 2.10). Low vapour permeability (Fig. 2.6) is wanted.

Piping. The good corrosion-resistance of many plastics in the presence of water or ordinary liquid wastes from buildings makes them attractive for piping in buildings. They are beginning to be used rather widely for drains, vents, and waste lines, but some users hesitate to employ them for water, especially hot water, under pressure because of fear that thermoplastic pipe might gradually creep, bulge, or sag. Nevertheless, their use for water supply lines is also increasing (Fig. 2.11). Chemical plants commonly employ different plastics for piping to transport different solvents or corrosive liquids.

These are only a few examples, there are many others such as hardware and mouldings.

2.3. COMPOSITES

For many applications, plain unmodified materials are inadequate to meet demands such as strength and stiffness combined with minimum weight, resistance to environmental conditions, or other exacting requirements. Recourse must frequently be made to composite materials in which several constituents are combined to provide synergistic behaviour unattainable by the constituents acting alone. Such materials have become familiar in space vehicles and transport generally, and are undergoing intensive development. They are becoming increasingly common in building. Plastics play a part in many of them.

A precise definition of composites is not easy to achieve. At the atomic scale, materials are composites of subatomic particles; at the crystalline and molecular scale, they are composites of various atoms; and at the microscopic scale, many materials are composites of different constituents. The term 'composites' is usually applied to combinations in which the constituents retain their identities at a more gross macroscopic scale, but are constrained to act together.

Just as it is not easy to define composite materials precisely, so it is not easy to put composites into neat pigeonholes. A classification often used is fibrous, laminar, and particulate. In fibrous composites, high-strength, high-stiffness fibres are embedded in a continuous matrix; in laminar composites, layers of different materials are bonded together and may be impregnated by an adhesive or matrix; and in particulate composites, particles of different materials are embedded in a matrix. Further classifications sometimes proposed include flake, or flat particles embedded in a matrix; and skeletal, a skeleton of materials interpenetrated by and filled with a matrix.

In building, the most used of all materials is the particulate composite, concrete. Mineral aggregates are embedded in and bonded together by a matrix of Portland cement and water. Plastics are not normally employed, although, as is brought out later, they may be incorporated for special purposes in concrete and mortar.

This discussion will concentrate, principally, on fibrous and laminar composites.

2.3.1. Fibrous composites

For many structural and semi-structural applications, including building, unmodified plastics are too weak and lacking in stiffness. High-strength, high-stiffness fibres are, therefore, incorporated to increase both of these properties. The fibres, in turn, being fine and

P.C.S.—4

slender, are unable to hold any shape and must be held in position by a matrix that supports them against buckling and transmits stress from fibre to fibre in shear. Each, therefore, contributes to the combined action of the composite, which transcends the capabilities of fibre and matrix acting alone.

Fibres. Both natural and man-made fibres are employed, but the most commonly used today is glass. In massive form, glass is relatively weak in tension, but when drawn into fine filaments of the order of 3×10^{-4} in (8 μm) in diameter, the strength rises greatly. Laboratory-drawn glass filaments are reported to reach strengths greater than 10^6 lbf/in² (6900 MN/m²). Commercial glass filaments attain 4×10^5 lbf/in² (2800 MN/m²) to 7×10^5 lbf/in² (4800 MN/m²) and, even after degradation during fabrication into finished parts, have strengths in excess of $2 \cdot 5 \times 10^5$ lbf/in² (1700 MN/m²). On a strength-to-weight basis, it is one of the strongest of engineering materials (Table 2).

The stiffness of glass, as measured by its elastic modulus of 10^7 lbf/in² ($6 \cdot 9 \times 10^4$ MN/m²), is high compared with plastics, but modest in comparison with steel. In a composite, because only part of the total is fibre, and the fibres may be oriented in different directions, the over-all elastic modulus is lower and is often a seriously limiting factor in such applications as aircraft and spacecraft where the high strength-to-weight ratio is otherwise attractive. Intensive research is, therefore, underway to find low-density, high-modulus, high-strength fibres.

Among the most interesting materials for such high-performance fibres are carbon, graphite, boron, and beryllium, all of which are low in density and, theoretically at least, can provide high strength and elastic moduli, the latter several times as high as steel. Progress is being made in research and development with all of these and other attractive materials. These fibres are in limited production and are being employed in composites for special high-performance applications such as compressor blades in jet engines.

Still more promising, at least theoretically, are the whiskers, extremely fine single-crystal fibres of such materials as aluminium oxide, silicon carbide, and others. Because of their single-crystal form, these fibres can approach their theoretical strengths and elastic moduli. Strengths of 3×10^6 lbf/in² (21×10^3 MN/in²) accompanied by correspondingly high elastic moduli have been reported for laboratory-prepared materials. Some, such as silicon carbide, are in limited production as both fibre and flake (Table 2.2).

Table 2.2. *Properties of reinforcements*

Filament type	Density kg/m^3	Tensile strength 10^3 lbf/in^2 (MN/m^2)	Young's modulus 10^6 lbf/in^2 $(10^3$ $MN/m^2)$
Continuous fibrous reinforcements			
E glass	2550	500 (3450)	10·5 (72·4)
S glass	2500	650 (4480)	12·6 (86·8)
Carbon/graphite	1500	350 (2420)	30 (207)
Boron	2360	400 (2760)	55 (389)
Beryllium	1830	185 (1280)	44 (304)
Stainless steel	7740	275 (1900)	29 (200)
Discontinuous fibrous reinforcements (whiskers)			
Aluminium oxide	3960	3000 (20 700)	62 (427)
Silicon carbide	3180	3000 (20 700)	70 (482)
Graphite	1660	2845 (19 600)	102 (703)
Iron	7830	1900 (13 200)	29 (200)

Broutman, L. J. (1969). 'Mechanical behavior of fiber-reinforced plastics', Dietz, A. G. H. (ed.), *Composite engineering laminates* (MIT Press).

Matrices. The most commonly used matrices are plastics, but for some applications, metals such as aluminium or still higher-melting alloys are required.

Many plastics are employed. Among the most common are thermosets such as unsaturated polyesters, epoxies, phenol formaldehyde, and melamine formaldehyde; but thermoplastics of the vinyl group and others based upon styrene and its copolymers are similarly reinforced with fibres for a great variety of purposes. Many other thermosetting and thermoplastic resins find their usefulness enhanced by the addition of fibrous reinforcement.

Structural applications. For structural and semi-structural parts of buildings, glass-fibre-reinforced plastics, usually unsaturated polyesters, have certain advantages. Among them are:

    a. Formability. These materials have no inherent shapes and must be fabricated into the desired forms. These can be shapes such as shells, ribbed and corrugated slabs and sheets, and folded plates, doubly curved if necessary to provide maximum structural efficiency.

b. Strength and lightness. The high strength and relatively low density of glass fibres, combined with plastics, which are also low in density, provide strength and lightness. Strength can be tailored by orienting the fibres to meet the stress directions in the structural part.

c. Toughness. Because reinforced plastics are tough and impact-resistant, they can be thin, unlike some materials such as concrete whose minimum thickness and attendant weight is limited.

d. Light transmission. Thin reinforced plastics parts transmit a good deal of light. They can, therefore, provide combined enclosure, structural capacity, and light transmission.

But reinforced plastics also have limitations that must be circumvented for successful structural use. Among the principal ones are:

a. Low stiffness. Although glass has a relatively high elastic modulus, the combination of glass fibres and plastic matrix is likely to result in a relatively low over-all elastic modulus, especially when fibres are oriented in a random configuration, as in glass mat. Consequently, it is necessary to employ inherently rigid structural shapes such as shells to overcome the low stiffness of the material.

b. Cost. Compared with many other building materials, reinforced plastics are not cheap, and they must be stretched to the utmost to make them competitive. This is another reason for taking advantage of formability to provide inherently efficient shapes.

c. Uncertain durability. The relative newness of these materials means that they do not have a long history of proven durability. That record is being slowly accumulated, but it will take time to establish the favourable and unfavourable conditions.

d. Fire. Like all organic materials, reinforced plastics can be destroyed by fire. Formulation with incombustible fillers and chlorinated polyester, for example, can provide self-extinguishing materials that will not support their own combustion, but a hot enough fire will destroy them. They must, therefore, be employed and designed in ways that will not have disastrous consequences, just as other organic materials of construction must be used with circumspection.

2.3.2. Examples of plastics-based composites in building

From the foregoing, it follows that reinforced plastics are best used in inherently rigid, stable, formable shapes that utilize the capabilities of the material to the utmost in as thin sections as possible. This does not normally lead to standard structural shapes such as I-beams and angles or channels. It does lead to curved shells and to folded plates in which shape lends rigidity and stresses are such as to be resisted advantageously by the material, possibly by orienting the fibres in the best directions.

Moscow pavilions. An early example of curved shells was found in the pavilions fabricated in New York and shipped to Moscow for the American Exhibition held at the same time as the Soviet exhibition in New York (Fig. 2.12). These pavilions consisted of hexagonal canopies, 16 ft (4·88 m) across, on central columns or stalks 20 ft (6·10 m) high. Intended to be temporary, they were designed to with-stand 60 mile/h summertime storm winds. Each of the six petals of the hexagonal canopy was sharply doubly curved for stability and could be made only $\frac{1}{16}$ in (1·59 mm) thick, with $\frac{1}{4}$ in (6·35 mm) thick edge ribs. Column walls were also $\frac{1}{4}$ in (6·35 mm) thick. These thin sections were strong and stiff enough for the purpose, and the thin canopies transmitted a great deal of light.

Dome for Libya. A large dome for a sports building in Libya (Fig. 2.13a) has glass-fibre-reinforced plastics folded plates (Fig. 2.13b) moulded into large sections, in turn supported on the ribs and frames of the dome. Lightness of this semi-structural unit allows a reduction in the size of framing members and supports.

Roof shell-sheets. In this Italian development (Fig. 2.14), roof sheets of glass-fibre-reinforced polyesters are supported by central posts. At the support point, the sheets are thickened, strengthened, and gathered upward in ribs continuous with the sheet, thus demonstrating the ability to form and tailor these materials to meet the stress conditions.

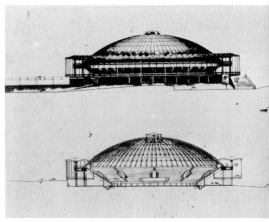

Fig. 2.13. (*a*) Diagram of 225-ft dome for sports building, Libya.

Fig. 2.14. Reinforced plastics roof sheets with formed and strengthened support points
Renzo Piano, Z. S. Makowski

Fig. 2.12. Reinforced plastics pavilion at American exhibition in Moscow

Fig. 2.13. (*b*) Folded plates of reinforced plastics for sports building
Mickleover Company, Z. S. Makowski

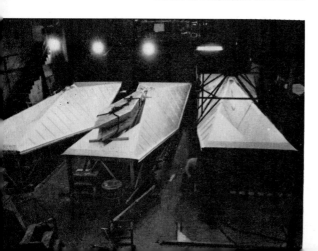

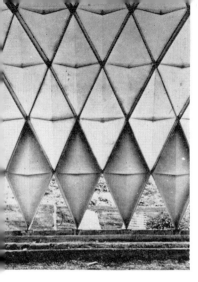

Fig. 2.15. (*a*) Folded plates of reinforced plastics forming vaulted roof.
(*b*) Tests of folded plates

Z. S. Makowski, B. S. Benjamin

Fig. 2.16. (*a*) Planar space frame of reinforced plastics pyramids combined with metal bars.
(*b*) Curved space frame of hexagonal reinforced plastics pyramids combined with metal bars

R. C. Gilkie, Z. S. Makowski

Folded plate vaults. By folding into various stiff configurations, the relatively low elastic moduli of reinforced plastics sheets can be largely overcome. The tendency of thin sheets to buckle under compressive stress is offset, as is shown in the portion of a roof-vault shown in Fig. 2.15a and in the test set-up illustrated in Fig. 2.15b.

Space structures. By combining hexagonal, square, rectangular, or triangular pyramids of reinforced plastics with bars of other materials such as wood or metal, it is possible to provide lightweight space structures combining support and enclosure. They may be planar (Fig. 2.16a) or curved (Fig. 2.16b) and, as is true of many such installations, they may also transmit a great deal of light.

2.3.3. Laminates

Sheet materials have been incorporated with plastics acting as impregnants and binders virtually from the first commercially successful use of phenol formaldehyde, the pioneer in this field.

Today, laminated plastics-based materials are widely employed as industrial and decorative materials. Sheet materials most commonly found are papers, woven fabrics, mats, wood veneers, and some metals. Papers include, among others, kraft and many printed decorative papers. Woven fabrics run the whole range from plain square weave to exotic patterns, and fibres include all the natural fibres such as cotton, hemp, sisal, and asbestos; and man-made fibres such as rayons, nylon, polyester, glass, and many more. Mats employ many of the same fibres. Thin wood veneers may be employed for engineering or decorative purposes. Metals may be facing sheets or interleaved for thermal or electrical purposes, the latter, for example, in printed circuits. Resins are mainly condensation-thermosetting phenol formaldehyde and melamine formaldehyde, and non-condensing epoxies and unsaturated polyesters, but other thermosetting and thermoplastic resins can be employed.

Sheet materials are usually impregnated with the plastic, assembled in piles of the required composition and thickness, and pressed. Condensation-type resins require temperatures in the vicinity of 300–50 °F (150–80 °C), and pressures ranging as high as 1 tonf/in² (15·4 MN/m²). Non-condensing resins require lower pressures and temperatures.

The final product may be industrial or decorative. Industrial materials are extensively employed in electrical and mechanical applications.

In buildings, decorative high-pressure laminates find wide use for counter tops, furniture and wall covering, especially where resistance to general wear and tear must be combined with appearance. The transparent melamine formaldehyde overlay and impregnant for the decorative surface sheet is resistant to wear and to water, alcohol, and other commonly found solvents. The back-up is kraft paper impregnated with phenolic resin. This thin laminate is in turn frequently bonded to a heavier substrate such as plywood or particle board (see below) for furniture, counters, and cabinet-work.

2.3.4. Structural sandwiches

Although structural sandwiches are laminar composite materials, their format and uses justify putting them in a separate category. In structural sandwiches, relatively thin facings of hard, stiff, strong, dense materials are bonded to relatively thick cores of lower density, weaker, softer, less stiff materials. The over-all structural behaviour is similar to that of an I-beam or box beam. In an I-beam, the flanges provide the resisting couple that opposes the bending moment of the imposed loads. In a structural sandwich, the facings perform the same function. In an I-beam the narrow web counteracts shear and provides the link between the flanges. In a structural sandwich the continuous core counteracts shear and provides the link between the flanges by means of an adhesive that bonds flanges and core together. The core also stabilizes the thin facings by preventing the wrinkling or buckling that would otherwise occur under compressive stress. The adhesive must be strong enough to resist shear and tension between the facings and the core (Fig. 2.17).

Sandwiches are widely employed in aircraft, spacecraft, and in transportation, generally, where their high ratio of strength and stiffness to weight is advantageous.

In buildings, sandwiches are also employed for structural and semi-structural purposes but must fulfill other objectives as well. Facings of building sandwiches may have to resist long-time weathering on the outside and normal wear and tear inside. Facings must also provide visual attributes such as colour and texture. Cores must

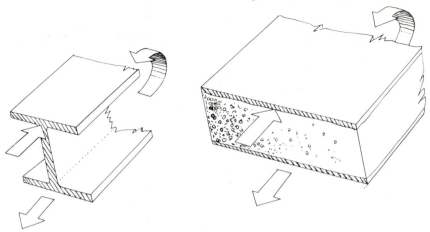

Fig. 2.17. Behaviour of I-beam and structural sandwich compared

provide thermal insulation and the entire sandwich may be called upon to provide acoustical isolation and a fire barrier.

High-rise flats for the Greater London Council exemplify the use of composite sandwiches to meet a number of requirements. The Council did not specify materials; it specified performance to be supplied by exterior wall panels. These were wind loads, over-all heat loss limits, acoustical attenuation, no flame spread on the surface, fire penetration resistance, minimum weight, minimum thickness, and minimum maintenance.

The resulting wall panels (Fig. 2.18) had moulded glass-fibre-reinforced, mineral-filled outside shells with a baked-on polyurethane finish, cores of wire-reinforced foamed concrete bonded to the outside shell with a flexible epoxy-urethane bond, and inner facings of wire-reinforced gypsum applied over a bitumen layer which bonded the gypsum to the core and provided a vapour barrier. The panels met all requirements, weighed one-sixth as much as standard wall construction, were one-third to one-half as thick, and could be preassembled to supporting steel and erected much faster than standard construction at a comparable in-place cost.

Another sandwich, employed in the cable-supported roofs of buildings at the Brussels and New York Fairs, in the translucent wall of the Apollo Assembly Building for the moon shots, and in smaller commercial and industrial buildings, consists of facings of thin,

Fig. 2.18. Wall panels for Greater London Council are composite sandwiches of reinforced plastics outer shell, foamed concrete core, and reinforced gypsum inner facing

translucent, glass-fibre-reinforced plastics bonded to a grid core of small aluminium extrusions (Fig. 2.19). The resulting sandwich, in addition to withstanding snow and wind loads, transmits a great deal of light. It therefore provides structure, enclosure, and light transmission.

Other facings include high-pressure laminates; rigid plastics such as vinyls and acrylics; cement–asbestos board; metals such as steel, aluminium, and titanium; plywood; hardboard; and thin reinforced concrete slabs. Cores include foams of plastics, glass, concrete, and calcium silicate; honeycombs of phenolic-impregnated kraft paper, reinforced plastics, and metal; a variety of grids; and miscellaneous other materials such as plywood and particleboard. Some are employed in buildings, others in transport and spacecraft or aircraft. In

Fig. 2.19. Sandwich panels having glass-fibre reinforced polyester facings bonded to aluminium grid core

Kalwall Company

precast concrete exterior wall slabs, for example, plastics foams such as polystyrene are commonly sandwiched between the thick structural slab and the thinner facing slab.

2.4. AUXILIARIES

Plastics are employed as auxiliaries to other materials. The latter perform the primary function, but the plastics auxiliaries are import-ant and may, indeed, be essential to make that function possible. Among the auxiliary uses are impregnants and additives, adhesives, sealants, and coatings. A few of such uses are briefly described below in connection with wood and masonry, together with a more general exposition of adhesives, sealants, and coatings.

2.4.1. Wood

One of the oldest materials, wood, is undergoing developments that enhance its natural properties, help to overcome its limitations, and extend its usefulness.

Impregnation. The beauty, toughness, and resilience of wood have long made it a favourite material for floors, furniture, and the trimming of buildings. For such purposes, protective and decorative coatings have customarily been employed, but these are surface materials only, subject to wear and damage, requiring renewal and repair, especially on floors exposed to traffic. Such finishes do not prevent, although they may retard, the swelling and shrinkage of wood caused by changes in moisture content that accompany fluctuations in the relative humidity of the surrounding air.

Numerous procedures have been tried to impregnate wood; that is, fill the fine porous structure with some material that would prevent the absorption of water or water vapour and would, at the same time, provide protection throughout the wood so that wear, for example, would merely expose more 'finish', thereby eliminating the need for refinishing. Oils, waxes, and other materials were tried by ancient craftsmen, with varying and only modest success.

The advent of synthetic resins and of new means of application show promise of fulfilling this old objective. The problem is to find a material of small enough molecular size to be able to find its way into the extremely fine structure of the wood cell and then to stabilize the material in place.

Acrylics are among the materials being developed for this purpose. Acrylics, like other synthetic resins, begin as small molecular aggregations, or monomers, that can be induced to agglomerate or polymerize into long chains. The monomer is a low-viscosity liquid, the polymer is a solid insoluble in many liquids such as water. The monomer can be forced into the wood by pressure, or vacuum followed by pressure, and finds its way into the structure of the cell. There it can be polymerized by the action of heat or, without raising the temperature, by high-energy radiation. Once polymerized, it is insoluble in most commonly found solvents, stabilizes the wood against dimensional changes, and provides an integral finish. By filling the pores of the wood, resistance to wear and damage is increased.

Other synthetic resins may be and are employed. During the Second World War, wood veners were impregnated with phenolic resins and laminated, practically fused, together under high pressure and heat. Under these conditions, the wood compressed approximately 50 per cent and became almost literally 'solid'. High strength and hardness were achieved, and the resulting 'compreg' was widely used for aircraft propellors. It is now employed for dies for forming metals such as aluminium sheet and for knife handles.

Impregnation has long been employed for protection against decay and fire. Materials such as pentachlorphenol, creosote, and various salts provide protection against decay and wood borers such as teredo (shipworm), termites, and various ants and grubs. Other salts, such as phosphates and those based on ammonium, can make the wood self-extinguishing in case of fire; that is, the wood will not support its own combustion.

Adhesively bonded wood. Glued wood has an ancient history. Several millenia ago, the Egyptians produced plywood and used some of the adhesives still employed today. With the advent of high-strength waterproof adhesives based on synthetic resins, glued wood construction has advanced rapidly. This has been particularly true of plywood, laminated wood, and particleboard.

In plywood, thin sheets of wood, or veneers, are glued together with the grain directions of adjacent layers, or plys, at right-angles to each other. The right-angled configuration is by far the most common, but special plywoods for boats and aircraft, for example, may be arranged in some other format. In laminated wood, layers of wood, usually boards or planks, are glued together with the grain of all members parallel. In particleboard, the wood is cut into chips, shavings, or other small pieces, coated with an adhesive, and pressed together in a random configuration into boards or other shapes.

Plywood more nearly equalizes strength in all directions and virtually eliminates shrinkage and swelling. Wood is strong and stiff in the direction of the grain but does not shrink and swell in that direction. It does shrink and swell across the grain but is weak in that direction. In plywood, the strength and stiffness in plys along the grain prevents shrinkage and swelling in the plys across the grain—the weak direc-

Fig. 2.20. Large laminated timber bonded with resorcinol formaldehyde
Timber Structures, Inc.

tion. In plywood, furthermore, splitting is virtually eliminated by the cross-ply construction. Large sheets can be manufactured, thereby facilitating the covering of large areas with strong, stiff material.

Laminated wood. Because thin boards can be bent easily to fairly sharp radii, and boards or planks can be joined end to end by a variety of interlocking or 'finger' joints, large sizes and curved shapes such as one-piece, long-span arches can be achieved that cannot be fabricated of saw timber. Moreover, the thin boards or planks are readily dried, whereas heavy timbers take a long time to season and are likely to be 'green' when installed. Consequently, large laminated members are dry to begin with and do not undergo shrinkage in place (Fig. 2.20). Different grades of lumber and species can be mixed, putting the high-strength, top-quality material only where it is needed and utilizing lower-quality lumber elsewhere. The unit cost of laminated timber, however, is higher than equivalent saw timber; consequently advantage must be taken of the efficient shapes and sizes available by laminating to reduce the amount of lumber needed.

Particleboard. Small pieces of wood, which would otherwise be wasted and burned, can be put through chippers or shavers and reduced to small particles. These can be coated with a thin layer of adhesive, such as urea-formaldehyde, for general-purpose material, and phenol-formaldehyde, for waterproof requirements, and formed under heat and pressure into boards or other shapes. The material may be made of all chips, or a board may have surface layers of flat shavings for appearance and strength. Some sawdust may be added. Because of the random configuration, strength is the same in all directions, and swelling and shrinkage are greatly reduced.

Exploded wood board. Another approach to the fabrication of boards is to saturate wood chips with high-pressure steam in an autoclave and suddenly release the pressure. The expanding steam explodes the wood fibres apart into a fluffy mass. This can then be pressed together again, using the natural lignin to bind the fibres. In the resulting 'hard board' the random configuration of the fibres results in equalized strength in all directions. Additives may be incorporated to resist attack by wood destroyers and to enhance other properties.

2.4.2. Masonry and concrete modifiers

Masonry. Brick and concrete blocks are generally laid up in mortars containing sand and binders, usually Portland cement or a natural hydraulic cement, lime, or both. When properly applied, such mortars provide joints strong in compression, moderately strong in shear, and weak in tension. Consequently, brick walls commonly must be designed to avoid tensile stresses, resulting in thick walls, usually not less than two brick widths in depth, with brick laid in various crossed patterns or bonds. If brick is used only as facing, with other materials such as concrete block as back-up, the same principles apply.

If plastics-based materials such as copolymers of polyvinyl chloride and polyvinyledine chloride are added in the form of a latex to the mortar, its tensile strength and the strength of the bond to the brick or concrete block are greatly increased, making it possible to rely more heavily on the tensile strength of the masonry assemblage. Thinner walls can be employed. In one experimental house, bearing walls two storeys high are one brick thick. It is also possible to pre-lay wall panels and hoist them into place.

Still greater strength is achieved if epoxy adhesives are employed to bond the masonry units together. Because the materials are expensive and because adhesive joints are best made as thin as possible, it may be necessary to grind the brick or block to more exact dimensions than are usually found, or to manufacture them to close tolerances. In the case of concrete blocks, lengths are usually more accurate than heights; consequently, these are sometimes laid up with the long dimension vertical, and only the horizontal joints are epoxy-bonded. Vertical joints are left open, or covered by plastic or stucco.

Concrete is a hard, normally durable material, strong in compression, but weak in tension. Although resistant to many environments including outdoor and most marine exposures, it does deteriorate when exposed to other conditions such as some of the de-icing chemicals employed on roads, cycles of freezing and thawing, and spillage in processing plants. Scaling and cracking caused by de-icers and by freezing and thawing are commonly combated by air-entrainment. It has been found that the addition of latexes based upon vinyl copolymers, acrylics, and similar materials are also effective in reducing such deterioration. Tensile strength and impact resistance are also increased.

For resistance to attack in some processing plants, a topping for concrete slabs may be employed incorporating similar plastics-based latexes. Epoxies are also employed. These can be applied in relatively thin layers, unlike usual concrete topping, and are, therefore, useful for patching, levelling, and general repair.

Cracks in concrete are frequently difficult to repair because of their fineness, but may be deleterious if water can penetrate and rust reinforcing steel, or cause further deterioration by freezing. Techniques are being developed to inject epoxies into such cracks. Epoxies bond tenaciously to concrete and, because they shrink little upon hardening, exhibit little tendency to crack away from the surrounding concrete.

P.C.S.—5

2.4.3. Adhesives

Glue has been used for several millenia to bond a great variety of materials. Ancient man seems to have noted that if animal parts were cooked, the resulting broth would bond wood and other materials firmly together. The Egyptians used it, as well as the casein resulting from curdled milk. We still employ animal and casein glues, as well as a variety of vegetable-based glues, such as cassava (tapioca), soya-bean, and the starch available from various sources. Although strong, these are all food materials subject to attack by micro-organisms, and many are softened by water, heat, or both.

High strength, durability. High-strength, durable engineering adhesives awaited the advent of the synthetic resins, and the first of these adhesives appeared after the Second World War. It was found that phenol-formaldehyde, familiar in moulded electrical parts, knobs, and many other parts, as well as the binder in laminated sheets and boards for electrical applications, could be employed to bond wood with an exceptionally strong and completely waterproof (boiling) glue-line immune to attack by moulds, fungi, and animal organisms. Water-proof plywood for boats, outdoor building applications, aircraft, and many other uses became available.

Phenol-formaldehyde requires heat for curing and this is not always available. Consequently, resorcinol-formaldehyde was developed, more costly, but capable of curing at room temperatures into a strong waterproof glue-line. Large laminated ships' keels and structural arches, for example, can be fabricated.

The problem of getting heat into the glue-lines of thick assemblages can be solved by placing the assemblage in a high-frequency radio field. The field penetrates the assemblage and quickly generates heat in the glue-line.

Both phenol- and resorcinal-formaldehydes are dark. Where light colours or no colour are important, as in gluing thin wood face veneers on plywood, urea- and melamine-formaldehyde are employed.

Nature of adhesion. The high-strength engineering adhesives have been accompanied by greater insights into the nature of adhesion. Formerly, it was widely believed that adhesion consisted of the inter-locking of the adhesive with the porous structure of materials such as wood and paper. Much research has shown that high-strength bonds can be achieved with hard non-porous materials such as glass, metal, and smooth plastics. Although much still needs to be learned about the nature of adhesion, the indications are that surface energies of the substrate (or adherend) and the liquid adhesive before it hardens are of great importance. It is certain that the adhesive must at some stage be liquid and must wet the substrate, that is, come into intimate molecular contact with its surface, if good bond is to be achieved.

Research has produced a great variety of strong, durable adhesives. To the adhesives already mentioned should be added the epoxies (a family of materials that bond exceptionally well to a great variety of porous and non-porous materials), urethanes, and many others. Some retain their strengths at relatively high temperatures; others are useful at low temperatures, or under other extreme ambient conditions.

Many of the synthetic resin adhesives bond strongly to porous and non-porous materials but are hard and relatively brittle. Materials such as metals expand and contract with temperature changes and are likely to break the bond. By modifying the adhesive with elastomers such as rubber and the softer more extensible synthetic resins, enough resilience can be imparted to allow for dimensional changes.

Today adhesives are employed in many critical applications, often providing joints stronger and more rigid than can be obtained with rivets, screws, and other mechanical fastenings. Adhesives are found in high-performance aircraft, spacecraft, water and land transportation vehicles, building materials and components, and great numbers of consumer and industrial items. With better insights into the nature of adhesion, the applications should increase.

Much still needs to be done. Reliable simple means of determining the quality of a joint without tearing it apart are badly needed. Extreme care must be exercised in making adhesive joints to ensure good quality. Such care is usually not possible to achieve in the field where vagaries of the weather, contamination of surfaces, and the inability to produce the close tolerances needed for high quality generally preclude the use of adhesives.

2.4.4. Sealants

With the advent of industrialized building, that is, the trend toward shop manufacture of large components such as slabs and big boxes to be assembled in the field, has come the critical need for materials to seal the joints between such components against penetration by rain and wind.

With changes in temperature, moisture content, or both, components such as panels made of concrete or metal expand and contract, and the motion is concentrated at the joints, opening and closing as the components contract and expand. Furthermore, even in field-fabricated parts of buildings, such as walls and roofs, motion occurs and relieving joints must be provided to prevent cracks. Such joints must also be sealed.

Requirements. Successful sealants must be able to accommodate the opening and closing of joints repeatedly, cold and hot, wet and dry, without fracturing because of fatigue. They must maintain watertight and airtight contact with the adjacent component whether stone, metal, glass, wood, brick, concrete, or whatever the material may be. They should not stain materials such as marble; neither should they sag, creep, nor otherwise distort with age. They should be durable even though exposed to extreme solar radiation, extreme temperatures, freezing and thawing, and polluted atmosphere. Ordinary putties, used successfully for a long time to seal small lights of glass in window sashes, are not adequate for these conditions.

Types. A new family of materials, based upon rubber and synthetic resins, has appeared to help solve this critical problem. They may be separated into two principal classes:
1. *Caulking compounds:* soft materials that are extruded from a gun or are applied with a knife, spatula, etc., and subsequently become stiffer and tougher;
2. *Gaskets:* preformed strips or other shapes that are snapped, pushed, or otherwise pressed into place under compression, but do not appreciably change consistency.
Both varieties have their advantages and limitations.

Caulking compounds are based on a variety of synthetic materials including polysulphides, silicones, urethanes, epoxies, and butyl rubbers. They are formulated as two-component and one-component materials. Two-component materials depend upon a chemical raction between the two constituents to convert the originally soft malleable mix into a tough rubbery consistency in place. They must be mixed on the job just before using and must be placed before the chemical reaction converts them to the rubbery state. Quality control, especially of job-mixing, is essential. One-component caulking compounds do not require job-mixing; they depend upon reaction with moisture in the air to convert them to the tough, rubbery state.

Both varieties of caulking compounds depend upon tight adhesion to the surrounding component materials, because at times, especially in cold weather, they are stretched as the joint opens. Great care must therefore be taken to see that surfaces are clean and dry or adhesion will be erratic. This is not always easy to achieve under field conditions. Good workmanship is paramount.

Gaskets do not adhere to the surrounding components and must be compressed enough when installed to make sure that they will always be under compression, no matter how much the joint opens. Tolerances must, therefore, be more closely controlled than in caulked joints. On the other hand, there are no time-limits on mixing and application, and surface cleanliness, although important, is not so critical as with caulking compounds.

It cannot be said that either caulking compounds or gaskets have completely solved the problem of joints. Much still needs to be done.

2.4.5. Coatings

Polymeric materials are bringing about marked changes and improvements in protective and decorative coatings. The standard divisions of paint, varnish, lacquer, and enamel are becoming less distinct as new formulations appear, but are still useful for classification purposes.

Some of the greatest alterations have occurred in paints. These coatings are usually thought of as mixtures of hiding, colouring, and filler pigments in a 'drying oil' vehicle, with or without modifiers such as 'driers' and thinners. The advent of polymeric materials such as acrylics and various members of the vinyl family, have brought to the fore the latex paints. These are not new; latex paints based on casein and rubber have been employed for years, but their application has been greatly extended by the availability of the plastics. In latex paint, the polymer is dispersed, along with pigments, in water, to which some solvent may be added if greater concentration is wanted. The paint is spread on the surface to be coated and, as the water-based vehicle evaporates, the polymer forms a dry tough skin incorporating the pigments. Hardening may occur quickly, unlike the slow hardening of drying oils, and successive coats of paint can be applied within a short time. Because the vehicle is water-based, tools and brushes can be cleaned readily with water.

A problem with latex paints, as with all paints, is that the film must be impervious to rain or other liquid water but be vapour-permeable; otherwise, if vapour is trapped behind the film and tries to escape, it will cause blistering and peeling.

Synthetic resins are widely employed in varnishes and lacquers. In varnishes, they may take the place of the natural resins traditionally employed, they may supplant the natural drying oils, they may provide both ingredients, or they may be employed together with the natural resins and oils in varying proportions. In lacquers, the resin is dissolved in a volatile solvent and spread by brushing or spraying. As the solvent evaporates, the resin forms a tough adherent coating on the surface of the substrate. Depending upon the volatility of the solvent, drying may be extremely rapid, or slow enough to permit manual brushing. Cellulosics such as cellulose nitrate have long been employed in lacquers, but other plastics, especially the vinyls, are being added to the family.

Coatings may be air-hardening or require baking. They may be brushed, rolled, sprayed, or dip-coated. They can be applied to a wide variety of substrates. Many plastics may be employed. Among the most common are the phenoplasts and aminoplasts such as phenol- and melamine-formaldehyde, the acrylics, alkyds, epoxies, polyurethanes, cellulosics, and members of the vinyl family.

BIBLIOGRAPHY

BENJAMIN, B. S. (1969). *Structural design with plastics*. The Society of the Plastics Industry, Polymer Science and Engineering Series (New York: Van Nostrand–Reinhold).

BROUTMAN, L. J. (1969). 'Mechanical behavior of fiber-reinforced plastics', in Dietz, A. G. H. (ed.), *Composite engineering laminates* (Massachusetts Institute of Technology, Cambridge, Mass.: M.I.T. Press).

CAMPBELL, K. J., DAVIDSON, J. W., and DIETZ, A. G. H. (1968). 'Reinforced plastics in multi-storey building', *Proc. 22nd Annual Meeting, Reinforced Plastics Division, The Society of the Plastics Industry* (New York).

DIETZ, A. G. H. (1969). *Composite engineering laminates* (Massachusetts Institute of Technology, Cambridge, Mass.: M.I.T. Press).

—— (1965). *Composite materials.* Marburg Lecture (Philadelphia: American Society for Testing and Materials).

——, GOODY, M. E., HEGER, F. J., Jr., McGARRY, F. J., and WHITTIER, R. P. (1957). 'Engineering the plastics house of the future', *Mod. Plastics* (June–July).

—— (1970). *Plastics for architects and builders* (Massachusetts Institute of Technology, Cambridge, Mass.: M.I.T. Press).

DU CHATEAU, Stephane. 'Les tridimensionelles industrialisees', *Techniques et Architecture*, 5 Revue Bimestrielle, 30e Serie.

HEGER, F. J., CHAMBERS, R. E., and DIETZ, A. G. H. (1962). 'On the use of plastics and other composite materials for shell roof structure', *World Shell Conference, 1962* (Washington, D.C.: Building Research Advisory Board, National Academy of Sciences).

MAKOWSKI, Z. S. 'Plastics structures', *Systems building and design* (University of Surrey).

Modern plastics encyclopedia, 1969–1970 (1969). (New York: McGraw-Hill.)

PLANTEMA, F. J. (1966). *Sandwich construction* (New York: J. Wiley).

Plastics in Building. Proceedings of a Summer Session, MIT (1967). The Society of the Plastics Industry, 250 Park Avenue, New York.

'Plastics properties chart', *Architectural and Engineering News* (March 1970). (Pennsylvania: Chilton Publications, Bold-Cynmyd).

SONNEBORN, R. H. (1954). *Fiberglass reinforced plastics* (New York: Reinhold). chap. 9, 'Design theory of reinforced plastics', Dietz, A. G. H.

Structural potential of foam plastics for housing in underdeveloped area (1946). Architectural Research Laboratory, University of Michigan.

Synthetic resins in building construction (1967). Proceedings International Symposium, Paris, 4–6 Sept. 1967. (Paris: Union Technique Interprofessionnelle des Federations Nationales du Batiment et des Travaux Publics).

'U.S. Pavilions in Moscow, 1969', *Mod. Plastics* (Nov. 1959).

ZERNING, J. *Structural forms with warped surfaces; their possibilities in plastics* (University of Surrey).

3

Weather as a factor in building design and construction

John K. Page, B A
Professor of Building Science,
University of Sheffield

3.1. INTRODUCTION

Everybody concerned with the construction industry knows that weather exerts an important influence, affecting both design of buildings, the actual construction process, and the amount of maintenance required. The majority of people in the construction industry treat weather as an accidental, though important, variable affecting the building industry for which really systematic planning is not feasible. For example, contracts are often phased optimistically in relation to weather. Subsequently, weather is presented to the client as a random, unpredictable factor justifying delays in programmes for clients. It does not always emerge clearly as a contingency against which contracts must be planned systematically. A mechanistic form of contract planning which takes the weather into account from day to day is clearly impossible due to the variety of weather met. But an operational form of contract planning which takes account of the statistical probabilities of encountering different combinations of weather is possible, and yet is seldom used at present. For example, on a short-term basis, far better use could be made of forecasts in planning day-to-day work.

Recent studies, for example the Construction Industry Research and Information Association (CIRIA) study on the effect of climate on earth-moving operations (1), have demonstrated clearly the improved planning which can result from systematic studies of the weather in relation to the building process.

This article attempts to review the various facets of the impact of weather on building design and the associated construction processes. It also discusses the durability problem briefly, i.e. the long-term performance in relation to weather factors. It covers a field to which relatively little research effort, directed specifically towards meteorological factors, has so far been placed, but there are signs of a significant expansion of effort in this area. For example, CIRIA has recently sponsored two studies in the University of Sheffield, Department of Structural and Civil Engineering; one to study the effects of weather on building production, and the other to study further the effects of weather on wet-weather earth moving.

3.2. THE PHYSICAL NATURE OF PROBLEMS ENCOUNTERED IN BUILDING CLIMATOLOGY: SOME BASIC DIFFICULTIES

3.2.1. Classification of fundamental physical problems

The problems dealt with in building climatology can be classified physically into four main classes:

1. Problems associated with momentum changes at the interfaces of the building, e.g. wind forces on buildings.

2. Problems associated with mass transfer both to building surfaces, and through building apertures, e.g. ventilation involving a mass flow of air, snow loading, rain deposition, roof drainage. Such problems usually involve wind and precipitation taken in conjunction.

3. Problems associated with energy transfer of one kind or another, e.g. heat loss and heat gain problems, thermal expansion, human thermal comfort, evaporation of moisture, daylighting illumination, photochemical degradation.

4. Problems associated with information transfer, e.g. forecasting of special hazards, location of climatological sensing devices for environmental control purposes, etc.

Many problems fall into classes that are mixed, for example, frost damage of porous materials involving mass transfer of moisture to produce the required water regime and energy losses to produce the damaging internal temperature regime, or state of ground which is dependent on rainfall, drainage, and the energy regime concerned with evaporation of the absorbed water in the soil. Working conditions on the ground are normally more favourable in summer, not because the rainfall is less but because the evaporative regime is more favourable.

3.2.2. Momentum transfer problems

Wind flow has been extensively studied in recent years as there have been a significant number of structural failures. Furthermore, tall building blocks in towns have perturbed the boundary layer flow and brought substantial quantities of air down to street level at high velocities which has led to serious complaints from pedestrians. There have been few significant advances in theoretical general predictive techniques in this area of design but the set of available experimental case-studies has substantially widened by recent extensive wind tunnel studies and also full-scale field studies. The approach remains a case-study approach. It is possible now to reach a reasonable understanding of the main features of the air flow round conventional building shapes. The primary meteorological difficulties relate to the single height sampling technique used in conventional meteorology to measure wind velocity and the consequent lack of information about vertical wind gradients. There is also lack of an adequate theory to deal with general topographic effects, and general difficulties with making wind measurements and estimates in urban areas due to the complex nature of the boundary layer flow. A representative extreme wind velocity in towns of rugged topography is much more difficult to obtain than most people realize. The author has studied a gale of extreme severity which damaged 125000 houses in his own city (2). Topographic problems presented formidable difficulties in assessing probable wind velocity in the vicinity of damaged buildings which could not be satisfactorily overcome. Wind failures are one of the few problems in building climatology where failure is due to a single meteorological factor. The only meteorological information needed is wind direction and velocity, which are fortunately usually measured simultaneously using an associated system of trace recording. However, the most interesting information for avoiding failure fortunately only occurs with great rarity. The ground-level wind-field in towns is so extremely complex that it is not easy to relate such flows to standard meteorological observations made at meteorological stations. Nor is it possible with the present state of scientific knowledge to set up general predictive theoretical models of aerodynamic flow.

3.2.3. Mass transfer problems

Traditional meteorology studies mass transfer to horizontal surfaces, using relatively simple methods of measurement—horizontal rainfall, rate of horizontal rainfall, horizontal snow depth. Building climatology is particularly concerned with mass transfers to vertical surfaces, and the wind-flow pattern exerts a dominant influence on what happens. For example, the concepts of driving rain and driving snow are of vital importance in water penetration and drainage studies. The driving rain reaching a vertical surface, however, is not uniformly deposited. A lot of the small drops go round the building in the air-stream, while a bigger proportion of the larger drops impinge, especially near the corners. As a result, an apparently simple meteorological parameter like 'rainfall on a vertical surface' cannot be scientifically defined as the water deposited depends on building shape and position of measurement on the vertical face of the building. The meteorological consideration of the problem must therefore be more complex than that involved in assessing undisturbed horizontal rainfall. Precipitation problems in building fundamentally involve the interaction between the gravity-field and air-flow patterns. Hence, distribution of water droplet size or snow-flake size, which will determine the vertical terminal velocity, must be considered as well as the intensity of mass flow on the horizontal plane and the associated wind velocity. Clearly it is possible to simplify to some degree by generalization, but conventional meteorology in compiling its records tends to separate the mass-flow data about rainfall on horizontal surfaces from the associated wind-flow data. Thus the wind speed and direction data is not normally summarized in association with rain and rate of rainfall. This has proved a fundamental difficulty in meteorological studies of rain penetration. Similar difficulties exist with snow. The use of computers is helping to remove this type of difficulty.

3.2.4. Energy transfer problems

Most progress in predictive studies in building climatology has so far been made in the sphere of energy transfer problems, partly for economic reasons concerned with the high cost of air-conditioning in buildings. These advances have been dependent on a thorough understanding of the heat transfer processes at the surface of building materials and of the thermal properties of solid and translucent materials. It has involved intensive studies of radiation climatology, especially in relation to vertical and inclined surfaces. Many advances in the practical working techniques for rapid estimation of radiation climates on slopes have been developed by building research workers. Since design usually involves an assessment of what happens on hot sunny days, clear sky conditions are usually assumed. The direct beam of the sun is fortunately not distorted by the urban form as the wind flow is, though it may be attenuated by the associated urban pollution. A systematic geometrical approach is possible. Estimation of diffuse radiation from the sky and ground is more complex, but high accuracy here is less essential as the direct radiation usually dominates on cloudless days. The outgoing radiation is usually less well considered. The radiation data has to be linked to other parameters affecting heat transfer, namely air temperature, wind velocity, and sometimes humidity in cases where latent heat transfer is involved. If the wind velocity is not known, it can be assumed to be suitably low, which gives a conservatively high surface temperature estimate. Fundamental physical difficulties have arisen from the fact that while the parameters actually required are usually recorded simultaneously, the classical methods of compiling meteorological statistics have led to the different climatic elements being listed separately in an unrelated way, because the problem of energy transfer was not clearly identified. It has proved a matter of considerable practical difficulty to get data which associates simultaneously air temperature, air movement, and humidity with direct and diffuse solar radiation measurements. At least, however, a tool exists in building radiation climatology that allows the mathematical prediction of the extreme values of the thermal response of a building system, even though the detailed calculations are, however, usually difficult, only becoming practical in complex cases by use of digital computing techniques.

An important consideration in building problems concerned with energy transfer is the rapidly varying nature of the transient energy transfer process. The diurnal pattern of the relevant meteorological energy parameters is thus of vital significance for predicting thermal conditions.

3.3. METEOROLOGICAL TECHNIQUES FOR DESCRIBING WEATHER

The previous section attempted to demonstrate briefly the complexities of the meteorological problems encountered in building climatology. The building industry must necessarily depend on the existing meteorological observations made by the network of stations which form the basis of any standard meteorological service. It is necessary, therefore, to consider the type of standard data available from such stations and any limitations in order to set building problems in a proper meteorological context.

3.3.1. Statistical versus real-time meteorological descriptions

Everybody who lives in the UK is only too well aware of the enormous variability from day to day of weather. One may talk about the weather last week, yesterday, or today. One may seek information about the future in the form of a forecast, for example, of the next day's weather using the aid of specialist meteorologists. All such descriptions are concerned with real-time descriptions, i.e. identification is made of the weather on a particular day at a particular time. However, as forecasts are essentially of short duration, even though they are likely in the near future to become statistically reliable over substantially longer periods than at present, say seven days, other kinds of description of weather are necessary for many building problems, which are necessarily concerned with periods of years, decades, or even centuries, far beyond the range of forecasting. One has, therefore, to invoke, to aid one's decision-making, the statistical techniques of climatology. Climatology attempts to provide a statistical description of the magnitude, relative frequency, and pattern in time of different meteorological parameters, taken either in isolation or in combination. It is necessary to have a set of records going back for a considerable time to compile reliable statistical summaries. In using such records for planning operations for the future, we basically assume that statistically the climate is invariant with time, i.e. one ignores climatic trends which from experience have a time-scale which is very long even compared, say, with the life of a building. Recent studies have shown there are distinct long-term climatic trends. There is some evidence that cold winters in the UK at present are statistically somewhat commoner than in the 1920s and 1930s. Such changes are normally taken into account in building climatology by using a relatively up-to-date set of climatological statistics derived from a recent selected period of thirty years.

An important start in attempting to use meteorological information, therefore, is to distinguish between real-time and climatological problems. Real-time meteorological problems are very rare in design but are very common in contracting. Contractors can clearly use both climatological data and forecasting techniques with advantage.

3.3.2. The nature of basic meteorological information used for forecasting and climatological purposes

Meteorological data is normally obtained from a network of meteorological stations using internationally standard instruments exposed in a standard way. The instruments are exposed in this standard way so that observations made at two sites are directly comparable, and systematic attempts are made to eliminate, as far as is possible, local effects, which would arise if instruments were exposed in different ways at different sites. Air temperatures, for example, are measured in the standard Stevenson screen with the thermometers mounted at a standard height in the standard, white-painted, well-insulated and ventilated enclosure. The shade air temperature at different parts of an adjacent building site, can be very different from those measured in the screen: for example, on a sunny windy day, the shade air temperatures in a basement excavation area sheltered from the wind could be appreciably higher. At the top of an eighteen-storey structure the shade air temperatures could be much lower during the sunny period of the day. On the other hand, the air temperatures at the top of such a building could be much higher than close to the ground on certain clear still nights. The concept of substantial micrometeorological variations from standard screen observations is thus very fundamental. Surface temperatures of objects in the sun differ widely from the observed shade air temperature.

Screen observations are normally made at fixed hours simultaneously over very wide geographical areas. Such simultaneous observations can be used for forecasting purposes by feeding them into a suitable meteorological forecasting system. They are also recorded and can be subsequently summarized in a systematic way. Such summaries provide the general climatological description of the meteorological site in question, and increasingly elaborate climatological descriptions become possible with advances in computer processing, provided climatological data is stored in ways suitable for high-speed data processing. The basic problem is to identify the best way to present such studies to aid the operational decisions in the field under

P.C.S.—6

consideration. This implies knowing precisely why the data is required and exactly how it will be used to aid decision-making.

3.4. CONCEPTS OF MACROCLIMATE, MESOCLIMATE, AND MICROCLIMATE

3.4.1. Macroclimate

Spatial differences of weather and climate across the country are of considerable interest to building designers and contractors. The substantial meteorological variations from place to place imply changes in the approaches required to give successful solutions. The broadcast level of description which might cover say the climate of Europe or the climate of the UK is known as the macroclimatic description. It is normally based on the data from standardized sites. A glance at the rainfall map for the Lake District by a contractor used to working in the south-east, is likely to show him that there is a need to plan more systematically in the hills against exposure and loss of time due to bad weather and rainfall in particular. The *Climatological atlas of the British Isles* provides a very useful compendium of general data of this kind (3) for the intercomparison of widely separated sites. Such maps provide information about the different weather elements at standard heights in a fairly broad scale. This is the macroclimatic scale of climatological description.

3.4.2. Mesoclimatic or local climate

If we reduce the geographical scale to cover an area of a few square kilometres, substantial variations in climate may be found; for example, sites on south slopes may be warmer than sites on north slopes. Cold air may drain into valleys and make them particularly cold on clear winter nights. Higher areas of ground may receive more rainfall than lower areas. Places very close to the coast may be windier than places a few kilometres inland. A great deal will depend on the topography. A description of the climate on this scale is known as the mesoclimatic scale. This scale of information is particularly important for town-planning purposes and for site selection. Any contractor who selects a site on a steep north-facing slope, exposed to the north-east winter winds, clearly gets off to a bad start on a high-class residential housing project compared with his rival who selects a sheltered, south-facing slope for his nearby competitive development. Thus systematic studies of mescoclimate can be most important in site selection, and the UK Meteorological Service now operates an advisory service which offers, for a modest fee, detailed guidance on the macroclimate of particular areas and adds information on the magnitude of any local variations which appear from the experience of the Office to be important.

3.4.3. Microclimate

Reducing scale again, one can discuss, say, the climate of the building site and the variations from one part of it to another. This scale of description is known as the micrometeorological scale covering variations over a scale of 10–100 m or so. This scale of variation can be very important in building design, for example, for considering the desirable location of outdoor living areas to optimize outdoor thermal comfort. It can also be important in contracting, for example, in selecting the best position for a crane on a windy exposed site so as to avoid wind channelled between closely spaced large building complexes and thus promote easy handling and aid operative safety.

3.4.4. Cryptoclimate

The smallest scale of all is the cryptoclimatic scale involving distances of less than 1 m, say, variations in the temperature across an upstand on a concrete flat roof. Knowledge of variations over this scale can help in the detailed interpretation of building failures due to climatic factors, for example, excessive differential thermal expansion.

3.4.5. Importance of vertical variations

Vertical spatial variations in climate must not be overlooked now that so many tall buildings are being erected. Reading progress reports on building contracts for tall structures demonstrates that many contractors tend, in particular, to underestimate the extent of interruption due to high winds. It is important to seek special information about vertical variations of climate when working on tall structures.

3.5. CONCEPT OF MICROCLIMATIC INDICES

In order to handle the practical problems encountered in building climatology, some method has to be evolved which allows proper consideration to be given to meso- and microclimatic factors in design and contruction. No standard meteorological compilation can hope to cover all the combinations of factors likely to be met with on the range of building sites at any one geographical location. It is possible, however, to easily obtain fairly reliable macrometeorological data for most areas in the UK. The problem is how to modify it to allow for meso- and microclimatic effects.

Geiger's classical book on microclimatology (4) was essentially founded on the concept that a general understanding of micro-meteorology could be derived from detailed study of a large number of well-chosen representative examples. His book fundamentally gives few actual predictive techniques to solve physically for the general case the particular problems mentioned. But, by providing enough examples, it enables the reader to estimate the direction of microclimatic modification in suitable cases with reasonable certainty using the technique of intercomparison, though there is the substantial difficulty of assessing the precise magnitude of any modifications in any particular situation.

It would appear that we have no choice in many aspects of building climatology but to follow the same path using case-studies. The problems of wind flow in urban areas, for example, are too complex, and the science of boundary layer fluid mechanics is insufficiently advanced to enable useful predictions to be made from theoretical mathematical models. Yet to use a wind tunnel, the only practical technique, one needs to produce a physical scale model of the urban complex and have sufficient time and money to carry out an investigation. This kind of process is far too slow and expensive for most design situations. So it seems that in many respects designers will have to continue to solve the problems they face by making predictions based on the published data of tests on a well-selected set of examples representative of contemporary urban aerodynamic phenomena. This approach has the limitation that it cannot solve radically new types of aerodynamic problems. Designers must be on their guard against such new aerodynamic situations, not adequately covered in the literature.

One way of handling the problem of local variations of climate systematically is by adopting the concept of microclimatic indices. A microclimatic index attempts to provide a generalized correction factor to be applied to standard meteorological data for a specific station in the network to enable mesoclimate and/or microclimatic considerations to be introduced on a systematic quantitative basis.

Two fundamental classes of microclimatic indices can be distinguished; local climatic indices and building microclimatic indices. Local climatic indices usually cover the mesoscale; for example, the effect of topography on exposure to wind and driving rain, or the effect of slopes on the solar radiation climate—the south slope being much more favourable than the north slope. They can also cover general urban influences. Building microclimatic indices, on the other hand, attempt to indicate the effect of particular building forms on their immediate environment. Examples of indices of this sort are provided by shelter diagrams for wind flow round buildings (5) where the consequent outdoor environmental conditions in the surrounding ground areas are dependent on the degree of shelter. The shelter can be stated by expressing the actual wind velocity in the sheltered area as a ratio to the free wind velocity. Another example is provided by tables of building surface temperatures expressed as a rise above the ambient air temperature. Such data can be linked to standard tabulations of extreme air temperatures to estimate extreme surface temperatures on particular surfaces.

The evolution of a systematic series of microclimatic indices for building purposes unfortunately is in its infancy and extensive scientific studies are still required to evolve a satisfactory series of indices to cover all aspects of building design. Another set may be needed for construction purposes. Some specific data on the sources of existing climatic indices for the UK is, however, included in the Appendix to this article.

3.6. THE USE OF METEOROLOGICAL DATA FOR DECISION-MAKING IN THE BUILDING PROCESS

The next problem to consider is how precisely to use the meteorological system to promote more effective decision-making in building. There are several fundamental divisions of decision-making in the building process, in which the application of systematic meteorological data is especially relevant.

3.6.1. Design against meteorological failure

The first field is design against meteorological failure. In this article a meteorological failure is considered to be any event which leads to a design objective not being achieved due to some meteorological cause. It is important for a start to distinguish between reversible and irreversible failures. The destruction of a cooling tower by a high wind is an example of an irreversible failure. A building being over-heated environmentally in a heat wave so that the occupants complain is a reversible failure. Cracking of a structure due to thermal expansion is an irreversible failure, while the temporary opening of a joint due to thermal expansion is a reversible failure.

Rational engineering design has always involved balancing the risk of failure due to under-design against the economic cost of avoiding failure. It is possible to reduce the risk of failure to very low limits by over-design, but there is an economic penalty for doing this which is often substantial. However, the possibility of failure is always present due to factors like unforeseen natural disasters, human shortcomings in engineering knowledge and construction achievement, and so on. The engineer's job is to make the probability of failure acceptably low. The decision about what is an acceptable risk can only be reached by hypothetically exploring, at the design stage, the various modes of failure in relation to the environmental factors likely to produce such failures using the best predictive techniques available that can be applied within the limited period of time and with the limited economic resources available for design. This involves a detailed understanding of the environmental forces, acting either singly or in combination, likely to produce failure. Modes of failure not anticipated and hence unexplored in design prove sometimes the undoing of the engineer and architect in practice. The acceptability of the risk of a particular pattern of failure must be related to the consequences of that failure. If the consequences are serious, a conservative safety factor may have to be accepted with the consequent economic penalties. If the consequences are trivial, a low safety factor can be used with the consequent reduction of construction costs.

Many building failures result suddenly from extreme climatic conditions, for example, those due to extremely high winds and heavy snow. Other failures are due to those long-term insidious effects of climate, which go under the heading of weathering. However, deterioration due to the weathering process also weakens the structure, so weathering has important long-term implications for structural safety. For example, corrosion of wall-ties of unsuitable material in cavity construction may eventually lead to 'floating' external walls, that become extremely hazardous in high winds. The author has examined a number of failures of this kind in Sheffield in late nineteenth-century housing where ungalvanized steel wall-ties were used embedded in black ash mortar. This led to serious corrosion of ties in the moist mortar of the external leaf and a large number of collapses occurred in the Sheffield gale of 1962 (2).

3.6.2. Dynamic versus static models of failure

Failure is obviously a dynamic process, but much building design proceeds on the assumption that dynamic loads may be replaced by suitable static loads which produce the same equivalent forces on the system. A particularly important class of problem, for which the static equivalent cannot be used, is the vibration problem; for example, oscillations of chimneys and galloping of power lines.

If the problem is of a dynamic nature involving accumulation of energy from cycle to cycle, the temporal pattern as well as the instantaneous magnitude of the meteorological inputs must be known in order to calculate the response of the system and the consequent stresses. Unfortunately, dynamic problems due to resonance effects can occur under relatively mild weather conditions if circumstances are favourable. For example, severe oscillations of the towers of the Forth road bridge were encountered in construction at relatively low wind velocities. This special class of climatological problem, therefore, cannot be handled by extreme value climatological analysis based on static equivalent loading. Unfortunately, existing meteorological records cannot provide the required detailed information for oscillating design, for example, the detailed vertical wind gust structure is important for tall mast oscillations but meteorological information in this kind of area is scant (6).

If the dynamic input is to be replaced by a statically equivalent meteorological variable, careful consideration has to be given to the basic response characteristics of the meteorological measuring instrument itself, and the time averaging subsequently used in the reduction of the meteorological data for building design use. The lower the inertia of any system, the quicker the response, and the shorter the appropriate averaging time. The latest approach to wind loading recognizes this facet of building climatology by using differing averaging periods for different aspects of wind failure. Similar problems can be encountered in other fields like the overheating of buildings. A thermally heavy building has a slow time response and consequently long averaging time may be used for the design meteorological data compared with a lightweight building that heats and cools quickly.

3.6.3. Destructive failures and extreme value analysis

The most important destructive failures are associated with wind and floods, either in isolation, or in combination. In some parts of the country, snow is an important factor. There were several structural failures due to snow loading in Yorkshire in 1962. Severe icing on tall guyed structures provides another cause of destructive failure. In the majority of cases of this sort there is a serious risk of loss of life and relatively high safety factors are mandatory, particularly for wind, snow, and ice.

Meteorological data for design must be carefully selected using statistically reliable data appropriately adjusted to allow for the effects of height and topography. Appropriate statistical techniques must be used to derive the design values (7). Engineers do not always appreciate that the extreme value recorded at a particular station is probablistically related to the length of the record. If two meteorologically similar stations have records of five years length and a hundred years length respectively, statistically one can expect a much higher extreme value for the station with the hundred-year record simply because it has had a far longer period in which high values could be observed. Systematic climatological design procedures require the reduction of all data from a network of stations with varying lengths of record to a common statistical base; for example, in wind design, basic climatological design maps for the UK have been prepared on a mean recurrence period of once in fifty years (8). If the life of the structure is very short, a somewhat lower value can be used. If the consequences of failure are very serious, for example, a nuclear power station, a longer mean recurrence period can be selected to give added safety. The selected wind velocity values at standard height have then to be adjusted to allow for actual height and topography.

The selection of the mean recurrence period to be used must be based on a combination of an analysis of the life of the structure and the consequences of failure.

For example, consider the problem of roof drainage where it is necessary to provide an adequate drainage system to carry away storm water safely. Two fundamental systems of drainage may be identified, internal and external. If external roof guttering overflows, it overpours to the surface below, and the consequences may not be very serious in the majority of cases. However, failure in the case of internal drainage can be very serious, leading to serious flooding of the building. It is thus normal to use different flow rates for design of the two roof drainage systems based on arguments of consequences. An external gutter system should normally be designed on the flow of 75 mm/h expected for 6 minutes once in five years, or 13 minutes once in twenty years, while for any system where overflowing cannot be tolerated 100 mm/h is used. This may occur for 3 minutes once in five years, and $6\frac{1}{2}$ minutes once in twenty years (9).

3.6.4. Combinations of meteorological loadings
Particularly acute difficulties arise from problems involving the combination of two or more weather variables, i.e. wind and ice, wind and snow. The designer has to explore systematically the various combinations likely to produce the worst design problem. The difficulty lies in the fact that the probability of one extreme being encountered in the presence of another is often very low, for example, heavy icing with an extreme gale. The combination of two extremes independently will produce an excessively conservative design, for example, the worst ice with the worst wind, but it is very difficult to decide on

permitted reductions in the present state of statistical knowledge about weather.

One may summarize the situation on design against destructive failure by advising designers to ensure both that they seek really reliable meteorological advice, and also that they understand beforehand the difficult statistical nature of the problems on which they seek such advice. A clear distinction must be made between oscillatory and non-oscillatory problems.

3.6.5. Weather and decision-making in environmental design

Environmental failure is defined here as the failure to achieve an acceptably comfortable indoor environment due to the impact of extreme outside weather on the building. It is important to distinguish between situations where the building fabric itself provides the main environmental regulatory control, for example, naturally ventilated buildings in summer, from situations where the internal environment is mechanically regulated by appropriate energy inputs or outputs to produce a stable comfortable environment, for example, heated buildings in winter and air-conditioned buildings with cooling in summer.

In the first class of problem, the only possible design approach is to optimize the internal climate by relating the properties of fabric to the known properties of the external climatic environment selected to match the conditions likely to cause extreme thermal discomfort. In many cases, it will not be possible to achieve comfort, but only minimize discomfort. The economic pay-off is difficult to assess for the output of the process is improved human comfort which is only indirectly expressed in terms of productivity. The meteorological design data selected, however, should be representative of the weather types causing most indoor discomfort, and humidity is often an important variable. This field is especially concerned with energy transfer problems, therefore the meteorological data must be in a form suitable to allow a proper assessment to be made of such energy transfer processes (10).

In the second class of problem where environmental control systems are available, the problem is to minimize the design and running costs of the environmental control system. In this case, there is a direct and assessable economic penalty for over-design, while, if there is under-design, the designer risks serious complaints about his competence from the users suffering discomfort. Users of this class of building assume they will be comfortable. Once again we are concerned with an energy transfer problem requiring detailed information on a temporal basis about meteorological inputs affecting the energy balance (11, 12).

Decision-making in environmental design has to be related to statistical information about the frequency of different types of relevant weather. However, the consequences of the failure are rather different in this field to the previous field discussed. Failure is expressed in terms of discomfort and increased running costs, together with complaints about the competence of design. Thus, while destructive failure has fortunately been comparatively rare in recent years, environmental failure has been extremely common.

The statistical approach to the environmental design problem is complex because of the large combination of weather elements affecting heat transfer, including solar radiation, long-wave radiation, wind velocity, air temperature, and vapour pressure. An added complication is the concept of a run of bad days, i.e. the concept of the heat wave or the cold spell. One day influences the next, thus indoor climate on one particular day is affected by the previous day's weather. Variations of energy inputs from hour to hour are liable to cause substantial difficulty in indoor temperature control in heavily glazed buildings.

Design in this field is often based on selecting a temperature design value which is only exceeded for a stated proportion of the total heating or cooling season. In some countries, for example, the design shade air temperature used for air-conditioning purposes has been the value exceeded on average for $2\frac{1}{2}$ per cent of the total cooling season. This approach does not deal with a run of days, and there is a statistical link between one hot day and the next. Complaints from users may therefore be still encountered with the simple statistical approach with the rare severe heatwave, due to the concentration of an extreme period into a run of days, one following the other for several days on end, which may not occur again for several years.

The selection of appropriate meteorological data is most important for air-conditioning applications, for here not only is initial design vitally affected, but also running costs. The climatic data used in this sphere of design is relatively sophisticated compared with most other fields of design (13).

3.6.6. Design against weather exclusion

Another important aspect of design is design against weather exclusion. This involves an understanding of the type, intensity, and frequency of weather to be excluded, for example, driving rain and its frequency, driving snow and its frequency, combined with a scientific understanding of the causes of failure, i.e. the combination of the aerodynamics of the impingement of precipitation and the aerodynamic forces causing the water to penetrate. Substantial advances have been made in this field over recent years (14). The concept of the driving rain index and the associated concept of the direction of the prevailing driving rain are important to many aspects of building design (15).

3.6.7. Failure of materials due to climatological causes

Two broad types of failure may be classified:

a. Mechanical and physical;

b. Chemical.

The field of weathering is a large subject and too big to explore in detail in a brief review. Mechanical and physical damage is usually associated with the problem of energy transfer which will determine the temperature regime and hence the consequent thermal movements. The energy regime is also linked with moisture movements. The concept of exposure is very important, and micrometeorological factors are of extreme significance. Frost is important especially following rain. The input of rain to various parts of the structure, together with the temperature and evaporative regime associated with these elements, will influence frost damage in different parts of building fabric. Precipitation is preferentially deposited along the top and edges of a building, and these areas are particularly liable to frost damage if unsuitable materials are used.

Chemical deterioration is also important. Such deterioration may be photoactinic; here the radiation climate of differently orientated surfaces plays an important part. Many plastics, paints, asphaltic materials, and bitumens are seriously affected by irradiation. The high energy photons in the ultra-violet and blue region of the solar spectrum play an important role so that photoactinic deterioration is more marked in places with a particularly clear atmosphere. The deterioration may, on the other hand, be electrochemical. The water retained in cracks and crevices helps accelerate corrosion due to electrolytic action, especially in polluted climates, where the rainfall may be very acid with a high concentration of electrolytic salts of one kind or another, which are retained in the crevices, producing a marked acceleration of corrosion rates in the presence of water (16).

It is difficult to generalize about the climatological design data required for decision-making in the weathering field. The causes of failure of a particular material must be known before the most relevant climatological data can be selected. Furthermore, the pattern of failure is often strongly linked to the cryptometeorological influence of the detailed design features. The concept of exposure has always been of vital significance in this field though it has not always been very precisely defined. Movement failures are usually associated with climatological extremes, for example, very hot weather, very cold weather, and once again one finds the most useful climatological data are extreme values of relevant weather variables. Photochemical deterioration, however, is a continous disintegrative process, and hence is more closely linked with mean value climatology, though few observations of ultra-violet radiation are available in the UK and assessments of exposure have to be made on the basis of over-all radiation inputs.

Particular difficulties are encountered with relatively untried new materials, and designers must carefully assess the probable weathering risks involved in the use of such materials. Study of field performance can be very valuable if intelligent appraisals are made of the degree of exposure of the inspected samples. It is always wise to seek out the adversely exposed areas in inspecting new building materials in the field, rather than be guided to sheltered and favourable sites by manufacturers' representatives. It is here that practical knowledge of microclimatic differences is so useful. Unfortunately, both human beings and materials favour a sheltered environment, and there is a natural human reaction against seeking out the more testing environments, especially in midwinter. It is necessary to distinguish between those materials likely to fail due to irradiation, those likely to fail from moisture, and those likely to fail from thermal stress in attempting to identify the adverse microclimates. Irradiation damage is often at a maximum on sheltered parts of buildings, facing the south and south-west sun, where relatively high surface temperatures may be encountered, which may help accelerate deterioration. Frost damage is associated with greater exposure to wind and water and is liable to be found in parapets and near the corners in the top part of multi-storey buildings.

3.7. OPERATIONAL ASPECTS OF BUILDING CLIMATOLOGY

3.7.1. General considerations

Planning against adverse weather is clearly an important aspect of construction planning. At the moment it is treated as more of an art than a scientific process on the majority of building sites. Some major contractors state that over-all hourly productivity varies by a factor of 10 per cent between summer and winter. The limited amount of daylight available in midwinter is certainly one of the most important factors setting back production, both at the beginning and the end of the day, but there are clearly a number of other factors.

3.7.2. Basic analysis

The starting-point in site planning must clearly be an analysis of the *possible* impact of weather on various aspects and stages of the contract. One can then proceed to analyse statistically the *probable* impact of weather on different operations at different seasons using climatological techniques. This implies a systematic analysis of the frequencies of various critical combinations of weather conditions likely to interrupt the different building processes. The main difficulty lies in deciding at what level of intensity of a meteorological event will be likely to interrupt a process. For example, how hard and for how long does it have to rain to produce a significant interruption of out-door site work? The answer is, of course, it depends on financial motivation, level of clothing protection, outdoor thermal conditions, site morale, management, and so on. On some sites, everybody dashes off into huts at the first sign of a splash of rain. On other sites, outdoor work that should be suspended in the interest of the job because of rain, continues unabated perhaps to meet some bonus award target.

For this kind of reason, there is a remarkable lack of information about the precise influence of weather on the various production processes, and contractors remain unsure about the actual data they need for improved planning.

3.7.3. Principal factors causing loss of production

The principal aspects of weather likely to cause significant losses in output are reduced hours of daylight in winter (made worse by the current trend to deeper building which implies less internal light for site work), heavy rainfall and snow, and low temperatures.

3.7.4. Low temperatures

Low temperatures affect productivity in a number of ways. There is, first of all, the problem of personnel protection against cold weather. This is not merely a question of low temperatures, but also a question of the cooling effect of the wind, i.e. wind speed. The cooling effect of the wind becomes even worse if clothing becomes damp, as additional evaporative cooling will occur causing severe chilling. The combination of low temperatures, high winds, and rain is particularly trying. The hands of operatives, even with protective clothing, are particularly vulnerable, especially as wet materials are handled. Such numbness reduces accuracy and affects productivity. Mackworth (17) for example, in investigating finger numbness showed the effects of raising the wind speed from still air to only a few metres per second could have as much effect as a fall of still air temperature of $5\,^{\circ}\mathrm{C}$. The problems of adverse weather conditions on tall structures derive more from air velocity effects than from vertical temperature differences, which are usually small in winter, especially with high wind velocities. Early closing-in of tall structures provides a valuable spur to productivity, and this should be a design aim in areas of high wind velocity.

3.7.5. Cold-weather working

Low temperatures are also important for a number of other reasons. The problems of frost protection against damage of new brickwork and stoppages of concreting in particular are liable to produce serious delays. It is, of course, possible to carry out concrete work at relatively low air temperatures if proper precautions are taken, and expertise in this direction has developed a long way especially in Scandinavian countries. This does not necessarily involve extensive preheating of aggregates and mixing water. Under mild conditions of night frost, the heat produced by the exothermic reactions of cement will be enough to confer protection, provided the work itself is sufficiently insulated to conserve heat and the air temperatures are not too low. Sheltered work is naturally better protected from excessive cooling than exposed work on tall buildings, where concreting from skips may present problems with excessively rapid cooling making it difficult to proceed in cold weather.

The problems of temperature conditions when casting in the ground are complex, but in the UK there is frequently a substantial store of thermal energy in the lower ground layers, especially in newly excavated ground early in the winter. Clearly, ground that has been exposed for some time will take up a temperature much closer to the air temperature. The present methods of control of concrete work, on a specification of air temperature conditions alone, are unnecessarily restrictive but there is little financial incentive to consultants who take the responsibility for safety to spend additional time on elaborating instructions to aid productivity with more enlightened technological control policies if only the contractor profits financially. If work stops at 3 °C everybody may feel absolutely safe, but the cost of such stoppages may be incurred unnecessarily.

3.7.6. Precipitation and the construction process

Three types of precipitation have to be considered on construction sites: rain, snow, and, in exposed locations, rime ice, which may be deposited directly from the clouds on to the structure. Such problems are encountered, for example, in high mast construction. The Building Research Station (BRS) studied building work on five sites in 1953–6, and the Station then observed that losses of man-hours, directly attributable to weather, amounted to from 0·8 to 1·8 per cent at the sites studied (18).

In the more exposed parts of the country and on tall structures these loss figures are likely to be excessively conservative. A later BRS study of the effect of the number of rain days per year and its wider influence on building operations suggest a total loss of 5·8 per cent with bad weather accounting directly for a loss of 1·3 per cent, reduced productivity estimated at 1 per cent, lost time at 1·3 per cent, and poor light at 2 per cent. Around London apparently it is raining for 5 per cent of the time, while this figure rises to 7–8 per cent in the western coasts of England and Wales, and in hilly districts in England to 10 per cent. In the north-west of Scotland the proportion reaches 15 per cent (19). Clearly some building operations are far more liable to severe interruption than others, for example roofing.

3.7.7. Rainfall types

Rainfall can be characterized in terms of total fall, duration of rainfall, intensity of rainfall, and associated wind speed and direction. An important practical distinction may be made between frontal rainfall and convective rainfall. Frontal rainfall typically consists of relatively small raindrops often carried along in a relatively strong wind. On exposed sites, such rain may be moving in a nearly horizontal direction, or even upwards when air is driven up over a building and is driven in through all unprotected vertical openings. Such rain may last for twenty-four hours or more, and may cause serious disruption on a building site, especially in winter when temperatures are low and drying is slow. Convective rainfall is caused by thermal up-currents in the atmosphere and is consequently much commoner in summer. Convective rainfall usually falls fairly rapidly with a relatively large drop size. The associated wind velocity is often relatively low, and the rain has a predominantly vertical direction. Such rain tends to produce shorter interruptions of building work, but may give rise to acute difficulties due to serious site flooding, and other drainage difficulties. In upland areas orographic rain, which may persist for very long periods, has to be planned against carefully.

3.7.8. Frontal rain

Persistent frontal rain is particularly common in winter, and is more intense on the western side of the UK, where the associated wind speeds are also greater than on the eastern side of the country, though there is an exposed belt along the north-east coast. Some general impression of the degree of site exposure likely to be encountered can be gained from the study of the number of rain days, i.e. days with precipitation of 0·2 mm or more. The BRS have suggested a threshold value of 0·5 mm of rain in an hour as the typical minimum amount of rainfall that is associated with actual stoppage of work on building sites. This may be a continuous fall at a rate of 0·5 mm/h, or may consist of one or more showers giving 0·5 mm in each hour. In Glasgow, for example, there is some rain in about 30 per cent of the hours in the year between 0800 and 1800 British Standard Time (BST), but rain with a rate of 0·5 mm/h can be expected for about 6 per cent of daylight hours (20). A special study in London for hours between 0700 and 1900 BST showed that the average monthly percentages of time likely to be lost due to appreciable precipitation range from under 3 per cent in April and June to nearly 6 per cent in February.

The BRS have had statistics prepared of the average number of hours in each month during which 0·5 mm or more rainfall (between the hours of 0800 and 1800 BST) falls for thirty stations in England (15).

A comparison of these statistics with those of annual rainfall for the same stations shows that the number of wet hours by day at a given place is approximately proportional to the annual rainfall, though there is some scatter in the results.

Table 3.1. *Relation between annual mean rainfall and annual mean number of wet hours (0·5 mm of rain or more in an hour) during working time (defined as 07–17 GMT). Data supplied by the Building Research Station.*

Annual mean rainfall		Annual mean number of wet hours
in	*mm*	*(percentage of time)*
20	500	3·6
25	625	4·3
30	750	4·9
35	875	5·6
40	1000	6·3
45	1125	7·0

The BRS map showing the driving rain index (combination of wind speed and annual rainfall) for the UK is also useful in revealing the relative degree of exposure of sites in different parts of the country. Revised data has been prepared by the Meteorological Office and will be published shortly.

3.7.9. Heavy rainfall

The heaviest downpours are normally associated with thunderstorms. These vary with frequency from place to place, for example, Glasgow has 8 storms per year compared with 17 in Birmingham and 16 in London (20). The risk of site stoppages due to flash flooding are thus heavier in the south-east of England, than in the west coast of Scotland where the main hazard is driving rain.

The frequency of rainfalls of a particular intensity can be estimated from Bilham's formula for intensities less than 25 mm/h (21).

$$n = \frac{121\,000t}{(r+2{\cdot}54)^{3{\cdot}55}}$$

where n = frequency (number of days per decade);
t = duration (h);
r = rainfall (mm).
For higher intensities Holland's modification should be used (22).

$$n = \frac{3820r \, \exp(1-0{\cdot}0315r/t)}{(r+2{\cdot}54)^{3{\cdot}55}} .$$

Variations in the UK are not large for peak rainfall intensities but there are some areas which are especially vulnerable, for example, the Lake District and parts of Devon.

3.7.10. Snow

Snow can also produce serious interruptions of building work. The snow-flakes, having a low apparent density due to the large amount of air included, have a low vertical velocity. The wind thus easily carries them along and snow driving into construction work can cause serious difficulties. There is a shortage of systematic data about effects of snowfall on construction sites in the UK but it is important to stress the vast differences in snowfall amounts and duration that can be expected in different parts of the country. It is particularly important to collect systematic snow data, when working in upland and mountainous areas.

3.7.11. State of ground

The state of ground depends on a number of factors other than just rainfall. Basically what happens depends on the balance between the input of precipitation and the removal of the deposited water by drainage and by evaporation. Evaporation plays a critical role in the balance, and during the winter evaporative losses may be very low. The soil becomes cumulatively wetter, especially if drainage is poor. Warmer weather associated with higher energy inputs from the sun and drier air associated with the spring winds from certain quarters rapidly accelerates evaporation and ground conditions begin to change rapidly because a more favourable balance is established. As the CIRIA study on wet location conditions showed (1), climatological data on solar radiation inputs, and evaporation characteristics, as well as studies of rainfall, have to be fed into any systematic study of the probable length of the earth-moving season.

Another hazard in cold weather is frozen ground, which may affect excavations. The depth to which the ground freezes is determined by:

1. The depression of air temperature below the freezing point of water;

2. The duration of sub-freezing temperatures;

3. The thermal properties of the ground and its cover (e.g. snow, grass, etc.).

Frost penetrates more easily in bare ground than in covered ground, as the vegetation provides some surface insulation. Ward (23) reports that the total number of days below 0 °C base temperature in a season provides a useful index (freezing index) of the depth to which the ground may be expected to freeze. In a severe winter spell with a freezing index of 133 (on a centigrade basis) the following depths of freezing were observed at Garston, Hertfordshire:

Well-drained sandy gravel	558 mm
Saturated silty sand	330 mm
Saturated clay	177 mm

The difficulty of excavating frozen ground varies with type of soil and degree of saturation at the time of freezing as well as the depth of frost penetration.

3.7.12. Daylight availability

The availability of daylight is clearly important for much construction work, but planning for daylight availability is often given insufficient priority. The availability of daylight decreases sharply in winter as one proceeds north in the UK, and sunrise occurs later as one moves west, as does sunset.

Table 3.2. *Times of sunrise and sunset (BST) at the solstices and equinoxes for every two degrees of longitude from 8° W. to 2° E. and for latitudes 56° N. and 52°N.*

A. 56° N.

Date		8° W.	6° W.	4° W.	2° W.	0°	2° E.	
21 March	sunrise	7 35	7 27	7 19	7 11	7 03	6 55	am
	sunset	7 46	7 38	7 30	7 22	7 14	7 06	pm
22 June	sunrise	4 45	4 37	4 29	4 21	4 13	4 05	am
	sunset	10 22	10 14	10 06	9 58	9 50	9 42	pm
23 Sept.	sunrise	7 17	7 09	7 01	6 53	6 45	6 37	am
	sunset	7 31	7 23	7 15	7 07	6 59	6 51	pm
22 Dec.	sunrise	10 02	9 54	9 46	9 38	9 30	9 22	am
	sunset	4 59	4 51	4 43	4 35	4 27	4 19	pm

B. 52° N.

Date		8° W.	6° W.	4° W.	2° W.	0°	2° E.	
21 March	sunrise	7 35	7 27	7 19	7 11	7 03	6 55	am
	sunset	7 45	7 37	7 29	7 21	7 13	7 05	pm
22 June	sunrise	5 11	5 03	4 55	4 47	4 39	4 31	am
	sunset	9 56	9 48	9 40	9 32	9 24	9 16	pm
23 Sept.	sunrise	7 18	7 10	7 02	6 54	6 46	6 38	am
	sunset	7 31	7 23	7 15	7 07	6 59	6 51	pm
22 Dec.	sunrise	9 38	9 30	9 22	9 14	9 06	8 58	am
	sunset	5 22	5 14	5 06	4 58	4 50	4 42	pm

In construction planning, one has not only to consider the availability of daylight on any construction site, but also the reductions of availability of daylight due to obstructions surrounding the site.

Detailed information about the availability of daylight may be found in a number of standard texts (24, 25). The present trend towards deep buildings which mainly utilize artificial light produces new lighting problems in construction that need proper consideration, as the consequent internal daylight factors are so low compared with traditional shallow building construction. In the absence of proper provision of artificial lighting, productivity may suffer seriously in such deep buildings. Further difficulties have been encountered due to the recent time system change to BST.

3.7.13. High winds and site stoppages
High winds are a particularly important cause of lost time on tall buildings, and on exposed sites a lot of cranage time may be lost.

There is also a serious risk of accidents if working is continued under severe wind conditions. It is important to recognize that the force of the wind depends on the square of the wind velocity, and marked vertical gradients of wind exist. Cranes attached to buildings need to be adequately secured to resist wind flow along the face of the building, as well as at right-angles to the face, for clearly in a closed-in building, the external air must move parallel to the façade close to the surface.

In assessing the probable interruptions to work due to wind, proper allowance must be made for increases in wind velocity as one moves up a tall structure. Very often a prevailing strong wind direction can be identified from climatological records, and the cranes used for the erection of the building should be placed, when possible, to gain the maximum shelter from the actual building under construction or from adjacent buildings. Particular care should be taken when tall slab blocks converge in plan to form V-shaped wind funnels. Avoid siting a crane in any consequent narrow gap, as the wind velocity may be substantially augmented.

The contractor must also bear in mind the responsibility he has for the stability of works during construction. Unless sufficient care is exercised, wind bracing and component attachment may be inadequate at certain stages in the construction process to withstand high wind velocities.

For lower buildings, forecast wind velocities at the 10 m level of 11 m/s or more are likely to produce significant loss of cranage time. Such velocities, for example, are likely to be exceeded at Glasgow Renfrew for 6·2 per cent of the time between 0800 and 1800 BST in January and 1·3 per cent in July. The mean value for the year is 3·4 per cent (20). When considering taller buildings, suitable allowances must be made for the substantial increase of wind velocity with height. S_2 factors to correct 10 m wind velocities to different heights given in *BRS Digest*, no. 199 (26) can help give some impression of the magnitude of this effect in the absence of aerodynamic disturbance, but detailed aerodynamic factors are likely to produce substantial modifications. Particular difficulties are likely in the space between tall buildings with low buildings upwind, and BRS studies of environmental wind show that wind velocities in the vortex zone between two such buildings can sometimes be double the free wind velocity

(27). This reinforces the argument for careful consideration of aerodynamic factors in locating cranes.

3.7.14. Forecast services and building

Forecast services are clearly important to the building industry (28). A wide range of forecast information is available from the Meteorological Office. Special forecasts can be provided for specific sites either on a routine day-to-day basis or to warn of specific hazards likely to cause severe interruption of work and problems of safety. Warnings of high winds and severe frost, for example, can be provided by telegram if required. Builders can also exploit the general forecast service profitably. The accuracy of forecasting is steadily improving as the techniques available for forecasting become more powerful. The use of the largest digital computer in the world will shortly enable the Meteorological Office to make substantial advances in forecasting accuracy and it is anticipated that it will be soon possible to forecast a week ahead with a fair level of statistical reliability. The Meteorological Office hopes to be in a position to forecast rainfall amounts in the near future and it is clear that the building industry will need to consider carefully how to organize to make effective use of the advances in meteorological forecasting in operational control of sites.

3.8. Conclusion

This article has surveyed the wide field of problems covered in building climatology, and has indicated the ways meteorological data can be used to promote better design decisions and improved planning of contracting processes. A great deal of research work is still needed to advance the subject further, but it is clear that the potential economic gains that could result from such studies are substantial and the field needs to be developed further by more detailed studies in a number of areas. Smith (29) reports that the losses due to adverse weather in an average year are £54 million, and in the exceptionally cold winter of 1962/3 rose to £180 million, which was roughly 8 per cent of new construction placed during that year (30). These merely represent the contracting losses. The economic losses due to ineffective building design, unnecessary weathering losses, and heavy services running costs have to be added. The economic advantages to be gained by the building industry, making more effective use of the techniques of building climatology, are thus very substantial.

The Meteorological Office are able to give a lot of detailed help, provided inquirers have a clear idea of the nature of the problems faced in building climatology. This article has been written on the basis of expounding principles rather than providing detailed data. The Meteorological Office keeps the national data bank in this field and should be consulted for detailed information for specific sites.

ACKNOWLEDGEMENTS

The author would like to acknowledge the extensive help he has received as a result of contact and discussion extending over many years with Mr Ralph Lacy of the Building Research Station and with the author's many friends in the Meteorological Office, especially H. C. Shellard and N. C. Helliwell.

REFERENCES*

1. NORMAN, R. (1965). 'The effect of wet weather on the construction of earthworks', *CERA Research Report*, no. 3.

2. PAGE, J. K. 'Field investigations of wind failures in the Sheffield Gales of 1962', *Proc. Symp. Wind Effects on Buildings and Structures*, Loughborough University of Technology, Paper 12, 12.1–12.23.

3. METEOROLOGICAL OFFICE (1952). *Climatological atlas of the British Isles*, M.O.488 (London: HMSO).

4. GEIGER, R. (1959). *The climate near the ground*, chap. 21, p. 215–29 (Harvard University Press).

5. JENSEN, M., and FRANCK, M. (1963). *Model scale tests in turbulent wind*, Part I (in English) (Copenhagen: Danish Technical Press).

6. DAVENPORT, A. G. (1965). 'The relationship of wind structure to wind loading', *Proc. N.P.L. Symp. No. 16, Wind Effects on Buildings and Structures*, **1**, 54–111 (London: HMSO).

7. THOM, H. C. S. (1966). 'Some methods of climatological analysis', World Meteorological Organisation, *WMO Technical Note* no. 81.

8. SHELLARD, H. C. (1965). 'The estimation of design wind speeds', *Proc. N.P.L. Symp. No. 16, Wind Effects on Buildings and Structures*, **1**, 30–51 (London: HMSO).

9. BUILDING RESEARCH STATION (1969). 'Roof drainage', *BRS Digest*, no. 107 (Second series) (London: HMSO).

10. LOUDEN, A. G. (1970). 'Summertime temperatures in buildings without air conditioning', *IHVE Journal*, **37**, 280–1.

11. MILBANK, N. O. (1968). 'Energy consumption and cost in two large air conditioned buildings', A paper presented to the *IHVE/BRS Symposium, Thermal Environment in Modern Buildings*, 1968, *BRS Current Paper* 40/68.
12. —— and HARRINGTON-LYNN, J. (1969). 'Estimation of air conditioning loads', *Air conditioning systems design for buildings*, (Elsevier).
13. INSTITUTE OF HEATING AND VENTILATION ENGINEERS (1965). *Guide to current practice* (London: IHVE) (revised SI version to appear in 1970).
14. CIB Report no. 51B (1967). *Report of the Symposium on Weathertight Joints for Walls*. Norwegian Building Research Institute, Oslo (Rotterdam: CIB, Bouwcentrum).
15. BUILDING RESEARCH STATION (1962). 'An index of exposure to driving rain', *BRS Digest*, no. 23 (Second Series) (London: HMSO).
16. FANCUTT, F., and HUDSON, J. C. (1960). 'The choice of protective systems for structural steelwork', *Proc. Inst. of Civil Engineers*, **17**, 405–30.
17. MACKWORTH, N. H. (1953). 'Finger numbness in very cold winds', *J. Appl. Phys.* **5**, 533–43.
18. MOPBW COMMITTEE ON WINTER BUILDING (1963). *Winter building* (London: HMSO) (contains a useful bibliography).
19. LACY, R. Private communication from BRS.
20. BRAZELL, J. M. (1970). 'Climatological information for the building industry, Glasgow', *WMO/WHO Symposium on Urban Climates and Building Climatology*. *WMO Technical Note* no. 109, 'Building Climatology', p. 223–32.
21. BILHAM, E. G. (1935). 'Classification of heavy falls of rain in short periods', *British rainfall* (London: HMSO).
22. HOLLAND, D. J. 'Rain intensity frequency relationships in Britain', *Hydrological Memorandum*, no. 33, also in *British rainfall* (1961).
23. WARD, W. H. (1963). 'Depth of ground freezing'. *Winter building*, Appendix 1, p. 23 (London: HMSO).
24. ILLUMINATING ENGINEERING SOCIETY (1962). 'Lighting during daylight hours', *IES Technical Report*, no. 4 (London).
25. HOPKINSON, R. G., PETHERBRIDGE, P., and LONGMORE, J. (1966). *Daylighting* (London: Heinemann).
26. BUILDING RESEARCH STATION. 'Wind loading on buildings', *BRS Digest*, no. 199 (Second Series) (London: HMSO).
27. WISE, A. F. E. (to be published). 'Effects due to groups of buildings', *Symposium on Architectural Aerodynamics* (Royal Society).
28. MEADE, P. J. (1966). 'Meteorological Office aims at tailor made site forecasts', *Construction News* (13 Oct.).
29. SMITH, J. R. (1969). 'Maximum efficiency in building in bad weather', *Building Technology and Management*, **7** (1), 4–6.
30. SIDWELL, N. (1970). 'The cost of private house building in Scotland', para. 113–7. Scottish Housing Advisory Committee. (Edinburgh: HMSO).

*For general summary of the whole field see *WMO Technical Note* no. 109, 'Building Climatology' and *WMO Technical Note* no. 108, 'Urban Climates'. (Geneva: WMO, 1970.)

Appendix

SELECTED SOURCES OF DATA ON LOCAL CLIMATE, MICROCLIMATE AND BUILDING CLIMATOLOGY IN THE UNITED KINGDOM

General

The *Climatological Atlas of the British Isles* (HMSO for the Meteorological Office), is a particularly useful general source of information.

Local climate

Topographical effects have been summarized by H. C. Shellard in an article on 'Microclimate and housing', *Architects' Journal*, **141** (1965), 21–6, while orientation effects are considered in the second article under the same general title in the *Architects' Journal*, **141** (1965), 81–4.

Urban climate

This is an extensive field of study. The most useful general book specifically associated with the UK is by T. J. Chandler, *The climate of London* (London: Hutchinson, 1965). The papers of the *Symposium on Urban Climates and Building Climatology*, held in Brussels in October 1968, which are now published in an abbreviated form by WMO/WHO and provide a useful summary of world knowledge in this field. (*WMO Technical Note* no. 108, 'Urban Climates'.)

Winter building

The MOPBW publication *Winter building* (London: HMSO, 1963), attempts to summarize data relevant to cold-weather working. Limited discussion of temperature and frost, humidity, rainfall, snow and hail, and wind data are included. Table 1 on p. 6 provides a useful check-list of weather factors against which winter precautions have to be taken. A useful summary of the present position may be found in two articles by J. R. Smith, Adviser on Winter Building, Ministry of Public Building and Works, published in *Building Technology and Management*, **7** (1969), 4–6, and 32–6. Table 3 gives useful quantitative guidance on the precautions to be taken in concreting at various shade air temperatures.

Wind general

H. C. Shellard has assembled general information about wind in the following publication, *Tables of surface wind speed, and direction over the U.K.* (London: HMSO, 1968).

Structural considerations—extreme velocities

BRS Digest, no. 99 (Second Series), provides a convenient summary of recent data. A map is provided of basic wind speed in metres per second of maximum velocity likely to exceed an average only once in fifty years at 10 m height. Elementary guidance provided on effects of topography. Tables provided to allow for variation of wind speed with height for terrain of different roughness.

The draft British Standard Code of Practice for Loading: *Wind loads* (Part 2 of the revision of CP 3, chap. v) provided similar information. Appendix C of the draft Code gave guidance on the effective height for loading purposes of a building near the edge of a cliff or escarpment. The new Code has just been issued.

A detailed review of the probem of estimation of design wind speeds may be found in the paper by H. C. Shellard, 'The estimation of design wind speeds', *Proceedings of the National Physical Laboratory Symposium No. 16, Wind Effects on Buildings and Structures* (London: HMSO, 1965). Also contains section on maximum wind speeds on mountain tops.

The problem of vertical gradients of wind velocity has been discussed by R. I. Harris, 'Measurement of wind structure at heights up to 598 ft above ground level', *Proceedings of the Symposium on Wind Effects on Buildings and Structures* (Loughborough University of Technology, 1968).

Environmental wind

The Building Research Station are carrying out extensive studies in this field. Some preliminary results of such studies were presented by A. F. E. Wise, D. E. Sexton, and M. S. T. Lillywhite, 'Air flow round buildings', *Urban Planning Research Symposium* (BRS, 1965).

The effect of shelter in the region of single buildings of various shapes is discussed in detail by M. Jensen, and M. Franck, *Model scale tests in turbulent wind*, Part I (in English) (Copenhagen: Danish Technical Press, 1963).

Wind flow in streets in Liverpool was studied by C. B. Wilson, and P. M. Jones, 'Wind flow in an *urban area*', *CIRIA Research Report*, no. 10, also reported in *Building Science*, 3 (1) (1968), 31–41.

A Symposium on Architectural Aerodynamics was organized by the Royal Society in February 1970, the proceedings of which will be published in due course. Recent BRS work on environmental wind was described at this meeting in a paper by A. F. E. Wise, 'Effects due to groups of buildings'.

Rainfall

Heavy rainfall and drainage. Heavy rainfall was discussed by E. G. Bilham, 'Classification of heavy falls of rain in short periods', *British rainfall* (1935). D. J. Holland showed the need for modifications in his paper, 'Rain intensity frequency relationships in Britain', *British rainfall* (1961), also available from the Meteorological Office, *Hydrological Memoranda*, no. 33. The implications for the design of urban sewer systems are discussed in *Road Research Technical Paper*, no. 55 while the design of building roof drainage is considered in *BRS Digest*, no. 107 (Second Series), 'Roof drainage'. The drainage of water from vertical façades is considered in *BRS Digest*, no. 23 (Second Series), 'An index of exposure to driving rain'. Frequency maps of daily rainfall of different intensities may be found in *Hydrological Memoranda*, no. 25. Local areas of very high extreme rates of fall were discussed by A. Bleasdale in a paper to the *Institution of Water Engineers*, 'The distribution of exceptionally heavy daily falls of rain in the United Kingdom, 1863 to 1960', **17** (1) (1963), 45–55.

Driving rain. This is discussed in *BRS Digest*, no. 23 (Second Series), which contains two valuable maps showing the relative severity of driving rain in different parts of the UK and the annual relative driving rain index for each of eight wind directions. Guidance is included on the effects of topography and building height. This data will be updated soon by new studies completed by the Meteorological Office. An up-to-date account of the consequent implications for design of weathertight joints may be found in CIB Report, no. 51B, *Report of the Symposium on Weathertight Joints for Walls* (Norwegian Building Research Institute, Oslo, 1967).

Snow. Some guidance on snow loading is given in British Standard CP 3, chap. 5, but this advice is fairly crude by current scientific standards. BRS are at present carrying out a systematic snow survey of roof loadings in the UK. Particularly useful scientific guidance on snow loading can be found in CIB Report no. 9, *On methods of load calculation, live loads, technological loads, snow* (CIB, Rotterdam, 1967). This includes an advanced scientific account of snow loading based on Russian and Canadian experience. This note takes proper account of aerodynamic factors, which are omitted from the existing UK code. A very useful summary in English of general Swedish experience in this field, including the economics of snow clearing appeared in a paper by R. Taesler, 'Problems caused by snow relating to building techniques', presented to the *Brussels Symposium on Urban Climates and Building Climatology. WMO Technical Note* no. 109, 'Building Climatology, p. 129–49.

Energy exchange with buildings

Temperature data. The standard source of temperature data is the Meteorological Office *Climatological Memorandum* no. 43 (monthly and annual maps of mean daily maximum and mean daily minimum temperature, and of average summer and winter temperature over Great Britain and Northern Ireland, 1931–60) (HMSO).

Extreme thermal exposure of building materials. Annual minimum and maximum shade air temperatures for more than a hundred stations throughout the British Isles have been published by L. P. Smith, 'Annual minimum and annual maximum temperatures', *Meteorological Magazine* (May 1965), 148–54. Maps of extreme temperatures are included in the *Climatological atlas of the British Isles*, MO 488 (London: HMSO, 1952). Such temperatures, of course, do not represent surface temperatures.

Thermal installation design. A recent general review of heating and air-conditioning aspects of meteorology is given by K. J. V. Fowler, 'Outdoor and indoor design conditions', *IHVE Journal*, **37** (1969), 73–81.

1. *Heating data.* The standard source of information in this field is the *Institution of Heating and Ventilating Engineers Guide*. The last edition appeared in 1965, and a metric revision is due in 1970. Included in the 1965 edition was data on cold days and cold periods in the UK, day-time and night-time temperatures, conditions of high wind and low temperatures. The *Guide* also includes typical records of severe conditions and information in degree days in the UK. More specialized information on cold days and cold spells may be found in *Post War Building Study*, no. 33, 'Continuously operated space heating installations, choice of basic design temperatures' (London: HMSO, 1965).

2. *Cooling data.* The *IHVE Guide* (1965 edn) also contains standard warm weather design data for the UK. A map is included of summer external design dry bulb isotherms for England, Scotland, and Wales. Summer external design dry bulb and design wet bulb data are tabulated for many countries in the world. Five UK locations are included.

BRS have been carrying out extensive studies of thermal conditions in buildings in summer, and have prepared systematic data for design purposes. The paper by A. G. Louden 'Summer time temperatures in buildings without air conditioning', *IHVE Journal*, **37** (1970), 280–1, summarizes their work in this area to date, and includes systematic information about solar radiation heat gains on horizontal and vertical surfaces. Daily mean outside shade air temperatures on sunny days for Garston, Herts., are tabulated, and frequency data on occurrence of hot days is included. A procedure is given for calculating consequent maximum indoor temperatures. BRS work on meteorological data for air-conditioned buildings is summarized in a paper by N. O. Milbank and J. Harrington-Lynn, 'Estimation of air conditioning loads in the U.K.', in Sherratt (ed.), *Air conditioning systems design for buildings* (Elsevier, 1969). Also see D. J. Nevrala, J. V. Robbie and D. Fitzgerald, 'A comparison of five digital computer programmes for calculating maximum air conditioning loads'. *HVRA Laboratory Report* no. 62.

Sun control (summer). Systematic design procedures for sun control have been published by P. Petherbridge, 'Sunpath diagrams and overlays for solar heat gain calculations', in *BRS Current Paper* Research Series, no. 39. Additional information can be found in the various papers published in the *Proceedings of the CIE Conference, Sunlight in Buildings,* 1965 (Rotterdam: Bouwcentrum International, 1967).

Typical temperature conditions in recently designed buildings in summer have been reported by F. J. Langden and A. G. Louden, 'Discomfort in schools from overheating in summer', *IHVE Journal*, **37**, 265–79, and F. J. Langden, 'Thermal conditions in modern offices', *RIBA Journal*, **73** (1965).

Sun provision (*winter*). The effect of slope on the sunshine climate is discussed by R. Geiger, 'Sunniness of different slopes', *The climate near the ground*, chap. 21 (Harvard University Press, 1959). H. C. Shellard provides energy data for the UK in his note in the *Architects' Journal*, **141** (1965), 81–4.

The requirements for the spacing of houses to meet the 1 hour minimum requirement for sunshine in living rooms is discussed in *Ministry of Housing and Local Government Planning Bulletin*, no. 5, 'Planning for daylight and sunlight' (London: HMSO 1964). The *Proceedings of the CIE Conference, Sunlight in Buildings*, mentioned above included several papers on the winter sunlight problem.

Daylight

The availabilities of different levels of daylight and the duration of daylight in the UK is considered in detail in the Illuminating Engineering Society *Technical Report*, no. 4, 'Lighting during daylight hours' (London: IES, 1962). Data are presented for Teddington/Kew, Eskdalemuir, and Lerwick. Such data has so far only been systematically used for design of daylighting of finished buildings. It could be used for the more scientific consideration of daylighting on building sites. Limited data on daylighting availability may also be found in *Winter building* (London: HMSO, 1963).

Atmospheric pollution

A general review of the atmospheric pollution problem in the UK may be found in Ministry of Technology Warren Springs Laboratory Report, *The investigation of atmospheric pollution, 1958–1966* (HMSO, 1967). A series of publications on a regional basis are in preparation. The Warren Springs Laboratory should be contacted for detailed information about specific areas in the UK. Urban aspects of atmospheric pollution are discussed at length in *WMO Technical Note* no. 108, 'Urban Climates', which gives a world wide review of critical issues.

4

The computer as an aid to architectural design: present and future

Dr Thomas W. Maver
*Director, Architecture and Building Aids Computer Unit,
University of Strathclyde, Glasgow*

4.5. Future developments
4.5.1. Government policy
4.5.2. Changes in professional practice
4.5.3. Integrated systems

Appendix 4.1. Selected bibliography of decision-making programs

Appendix 4.2. Selected bibliography of management programs

4.1. A MODEL OF THE DESIGN ACTIVITY

The effective application of computers to the architectural design activity has been hampered by the lack of a sufficiently general and valid model of the nature of the activity. It is proposed, in this opening section, to put forward two models; the first will allow a statement of design objectives to be made, the second will provide a framework in which present and future computer development can be located.

4.1.1. Cost-performance model

Markus (1) has described a cost-performance model developed by the Building Performance Research Unit in which the building-user interaction is described in terms of four subsystems: the building system, the environmental system, the user activity system, and the client objectives system. Fig. 4.1 illustrates how the client's objectives —production, stability, etc.—require, for their implementation, a number of activities to be undertaken—work-flow, communications, etc. At the other end of the model the building system—the fabric, services, and contents—give rise to an environment. There is two-way interaction between the environmental system and the activities system: the environment inhibits or promotes the activities to a greater or lesser degree and the activities, in turn, modify the environment. In any building, a balance is struck at this interface by modifying the environment (for example, jamming doors open, painting black-out on windows) or by modifying the activities (for example, one-way traffic in corridors, staggering lunch-hours).

Now the building system costs money to provide and the environment system costs money to maintain. Together these costs-in-use represent the client's investment in the building. At the same time a certain measure of performance is achieved in carrying out the activities and realizing the objectives. This performance represents the return accruing to the client. The designer's objective can now be identified as that of maximizing the return on the client's investment, i.e. providing a building which gives the best balance between cost and performance. Since cost is measured in pounds and performance

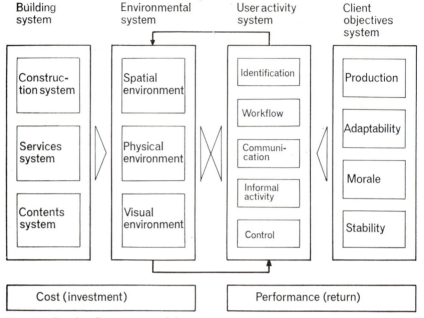

Fig. 4.1. Cost/performance model

may be measured in a large variety of disparate ways, the task is not an easy one. In other design professions a set of performance specifications are contained in the brief and the designer aims to satisfy them at minimum cost; in architecture, however, the opposite is generally the case, i.e. a cost limit is set and the architect aims to provide within that cost the best possible performance.

In any event an activity procedure is required which will maximize the designer's chance of achieving his objective.

4.1.2. Activity model

The design activity model shown in Fig. 4.2 is an adaptation of that proposed by Markus (2). Two dimensions are identified—the vertical one of design management and the horizontal one of design decision-making.

Design management is seen as a sequential progression from the general, strategic stages in design to the detailed, tactical stages; the

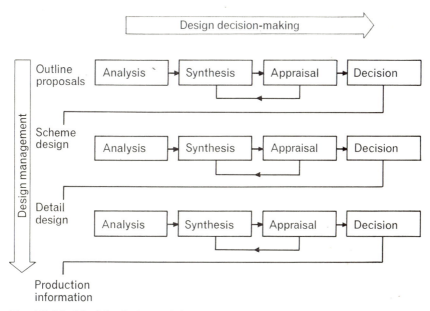

Fig. 4.2. Model of the design activity

titles given to the stages are in fact taken from the 'Plan-of-Work' set down in the RIBA Management Handbook (3). The management activities which must be undertaken include the handling of cost and performance information, the scheduling of the work force, job costing, communication with the client, etc., and a later section will cover programs applicable to some of these activities.

Design decision-making is seen as comprising three steps—analysis, synthesis, and appraisal—leading to the decision which is then carried forward to the next management stage. The definitions of the three steps are given as follows.

Analysis: the collation, ordering, and manipulation of basic data to promote the structuring of the problem or the establishment of basic relationships in the data.

Synthesis: the generation of partial or total solutions by the compounding of individual elements.

Appraisal: the measurement of the characteristics of a solution and its evaluation by comparision of the measures with criteria. A feed-back loop is shown between the appraisal step and the synthesis step to reflect the iterative nature of the decision-making activity; as will be shown in a later section the application of the computer to the decision making process allows many iterative cycles, and hence consideration of many more design alternatives than is possible when working in a purely manual mode.

4.2. THE COMPUTING MACHINE

4.2.1. Man/machine comparison

The relative attributions of man and machine are given in Table 4.1 which was drawn up by the late Paul Fitts and is reproduced from Singleton (4). It will be clear that the machine is designed not to emulate man but to complement him. Man is superior in the matter of: formulating general rules, laws, and relationships, particularly where the perception of space, depth, and pattern is involved; interpolation, extrapolation, and prediction; making judgements and decisions. The machine is superior in the matters of: inferring from general rules, laws, and relationships, a particular case; rapid computation; precise repetition; short-term storage of information.

4.2.2. Man/machine communication

The elements of man/machine communication are the program, the input and the output. The program is a set of instructions which detail, in logical order, the arithmetic processes which are to be performed on the specific data which is entered as 'input'. The results of the arithmetic processes are produced by the machine as 'output'. The efficiency of the machine relates to the fact that the program can be written once in a general form and used over and over again for different sets of input data.

In the last few years two major strides forward have been made in computer development—in language and in access facility—which are of great importance to the designer.

Language. The computer operates in 'machine language' which is made up of binary arithmetic notation, and in the early days of

Table 4.1. *Relative attributes of man and machine*

	Machine	Man
Speed	Much superior	Lag 1 s
Power	Consistent at any level. Large constant standard forces	2·0 hp for about 10 s 0·5 hp for a few minutes 0·2 hp for continous work over a day
Consistency	Ideal for routine repetition, precision	Not reliable: should be monitored by machine
Complex activities	Multi-channel	Single-channel
Memory	Best for literal reproduction and short-term storage	Large store, multiple access. Better for principles and strategies
Reasoning	Good deductive	Good inductive
Computation	Fast, accurate. Poor at error correction	Slow, subject to error. Good at error correction
Input sensitivity	Some outside human senses, e.g. radioactivity	Wide energy range (10^{12}) and variety of stimuli dealt with by one unit; e.g. eye deals with relative location, movement, and colour. Good at pattern detection. Can detect signals in high noise levels
	Can be designed to be insensitive to extraneous stimuli	Affected by heat, cold, noise, and vibration (exceeding known limits)
Overload reliability	Sudden breakdown	'Graceful degradation'
Intelligence	None	Can deal with unpredicted and unpredictable; can anticipate
Manipulative abilities	Specific	Great versatility

computing the programmer was obliged to communicate with the machine in these terms. It was soon realized that the machine itself could perform a translation function and this fact allowed symbolic languages to replace the machine language. Even the symbolic languages, however, were unwieldy and a third level of languages—

the compiler languages—came into being, of which Fortran and Algol are two examples. Recently, as computer usage has spread from a few specialist applications to the whole arena of industry, commerce, and education, a fourth generation of high-level, problem-orientated languages has been developed. It is now possible for the designer to program and operate a computer using commands such as Build, Divide Bay 33, Assign Activity 1 to Space 3, Calculate Cost, etc.

Access. In earlier years the choice facing the potential user was that of buying a computer or of sending data and results by post or carrier to a machine installed in another building, or even in another city. The first alternative, to all but the largest design practices, was economically unviable; the second alternative was anathema to effective man/machine interaction when iterative procedures were being implemented. Now, however, the concept of remote multi-access time-sharing is a reality and is offered to design practices by a number of commercial bureaux. In effect, a bureau rents a terminal for the user and sells him computing time by the minute. The terminal (a little larger but otherwise very similar to an electric typewriter), is sited on the user's premises and linked by ordinary GPO telephone line to the bureau's computer. By dialling the computer's telephone number, connection is made and to all intents and purposes each subscriber has exclusive use of the central machine. The program is written on the keyboard of the terminal and data is input via the keyboard or by perforated paper tape. The output is typed out by the terminal on a continuous roll of paper. Most bureaux offer the user access to standard library programs covering commonly recurring problem areas; otherwise the user may store his own programs, at a small cost, in the computer's files.

Costs for such a system are not high. The rental of terminal, GPO line, and modem (which connects the terminal to the telephone line) amounts to approximately £400 per annum and computing time is bought at about £8 per hour. Telephone charges are at normal STD rates.

A high-level language used on a time-sharing system gives the small and medium-sized practice an effective computing facility. The degree to which the system is usable by inexperienced personnel is clear from the example which appears later on information retrieval.

As yet the designer, when he is in a position to invest a very large sum of money, must communicate with the computer in alpha-numeric terms, i.e. in letters and numbers. It is anticipated, however, that before long graphic terminals, on which the designer can sketch alternative solutions, will be as readily available as are the keyboard terminals at present.

4.3. DECISION-MAKING PROCESSES

The three steps in the design decision-making activity—analysis, synthesis, and appraisal—have already been defined. In this section, examples of computer programs relevant to each will be described and a summary of other programs given.

4.3.1. Analysis

As defined, analysis is primarily concerned with the modelling of mathematical relationships between variables. Two mathematical processes—linear programming and cluster analysis—will be ex-emplified.

Table 4.2. *Description of residential blocks*

Type of block	Number of flats in each category				No. of storeys	Plan area (100 m²)	Cost per block (£000)
	1-room	*2-room*	*3-room*	*4-room*			
A	2	4	8	—	3	5	52
B	—	6	12	—	6	5	80
C	4	4	4	8	2	8	75
D	—	12	—	16	8	6	120
E	—	—	20	10	6	8	120

Linear programming. Linear programming (LP) is defined as the optimization of a linear objective function subject to a set of linear-constraints. Consider the following problem, taken from the work of Krejčiřik and Šipler (5). Table 4.2 gives details of five different types of standard housing block which may be used in a large housing development project. The planning task is to determine how many blocks of each type are to be utilized in the development such that:

   a. The cost is minimized;

   b. One thousand flats are provided;

   c. There are no fewer than 80 one-room flats;

   d. There are no fewer than 250 two-room flats;

   e. There are no fewer than 250 three-room flats;

   f. There are no fewer than 250 four-room flats;

   g. The average number of storeys in the development does not exceed 6;

   h. The plan area of the development does not exceed 300 units.

To find a feasible solution to this problem, let alone an optimum one (i.e. one which minimizes cost) by any other means than LP is a well-nigh impossible task. Since computer programs exist for the solution of LP problems, however, the designer's task is simply one of expressing the requirements *a–h* in linear equation form. Let x_A, x_B, x_C, x_D, and x_E be the number of each type of block. Then:

from *a* $52x_A + 80x_B + 75x_C + 120x_D + 120x_E = \text{min. cost}$

from *c* $2x_A + 4x_C \geqslant 80$

from *d* $4x_A + 6x_B + 4x_C + 12x_D \geqslant 250$

from *e* $8x_A + 12x_B + 4x_C + 20x_E \geqslant 250$

from *f* $8x_C + 16x_D + 10x_E \geqslant 250$

from *b* $14x_A + 18x_B + 20x_C + 28x_D + 30x_E = 1000$

from *g* $-3x_A - 4x_C + 2x_D \leqslant 0$

from *h* $5x_A + 5x_B + 8x_C + 6x_D + 8x_E \leqslant 300.$

The first equation represents the objective function in linear terms, the remaining equations and inequalities represent the constraints in linear terms. The procedure now is to input the coefficients of these equations and inequalities, run the program and obtain the results as output. In this case the output read:

$$x_A = 11$$
$$x_B = 9$$
$$x_C = 14$$
$$x_D = 14$$
$$x_E = 0$$

and the running time, on a commercial time-sharing system similar to that described in the previous section, was 20 s.

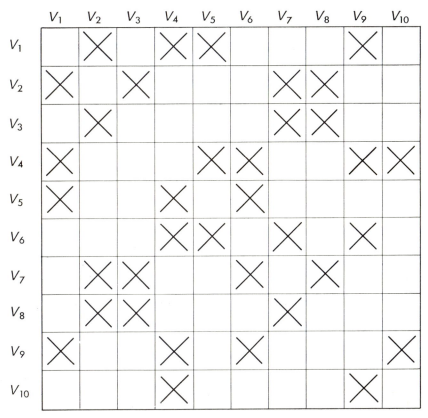

Fig. 4.3. Relationship matrix

Cluster analysis. Cluster analysis (6) is a technique for structuring a very large multi-variate problem into parts such that each part can reasonably be dealt with in isolation. Consider Fig. 4.3 which records the existence (by means of a cross) of a relationship between the variables v_1, v_2, v_3, v_4, etc. In this simple example it is possible to construct a network diagram of the matrix and identify two relatively discrete subsets (Fig. 4.4); in a very large problem, however, the task is extremely difficult and in order to solve it computer programs have been written. The basis of the program logic is that the subsets are defined either by maximizing the number of links within each

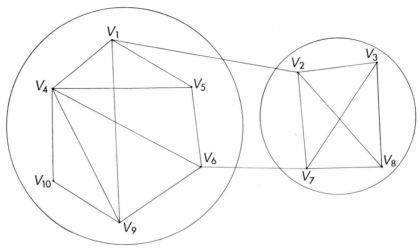

Fig. 4.4. Relationship clusters

subset or by minimizing the number of links between any subset and any other. Where the variables v_1, v_2, v_3, v_4, etc., are actual elements of space (for example, director's office, store-room, reception, etc.) the cluster solution expresses a physical grouping of the spaces.

4.3.2. Synthesis

It will be clear from the section 'man/machine comparison' that the process of synthesis is not one for which the computer is ideally suited since it is poor at pattern recognition and has little or no innovatory facility. None the less, those involved with the development of computer aided architectural design techniques have produced programs which attempt to generate built form layouts on the basis of a few obvious design criteria; an example of such a program is given in this section. First, however, an example will be given of a program which, in a limited sense, generates a solution to a pipe insulation problem by selecting from a given range of possible solutions the least expensive.

Pipe insulation. Consider Fig. 4.5 which shows the cross-sectional area of an insulated hot-water pipe. The problem is to select a thickness of insulation which, for a given fluid temperature, ambient

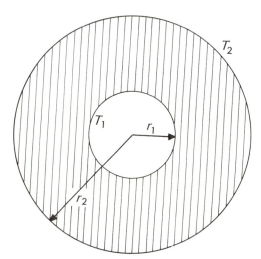

Fig. 4.5. Pipework insulation

temperature, cost per unit of heat generated, and pipe radius, is obtainable for minimum cost. The input is as follows: pipe radius r_1, fluid temperature T_1, cost of generating a unit of heat per hour C_H, the thermal conductivity of the material k, and the cost of the material per unit thickness. The programs may proceed as follows:

1. The capital cost of the insulation material is amortized over its life to give a per hour cost C_I;

2. The radius r is set to a numerical value r_2 and $\dot{Q}$ found from

$$\dot{Q}_{r_2} = 2\pi \frac{r_2 - r_1}{\log_e(r_2/r_1)} k \frac{T_2 - T_1}{r_2 - r_1}$$

where $\dot{Q}_{r_2}$ is the rate of heat loss from the material;

3. $\dot{Q}_{r_2} \times C_H$ is calculated;

4. $(r_2 - r_1) \times C_I$ is calculated;

5. $(\dot{Q}_{r_2} C_H) + (r_2 - r_1) C_I$ is calculated and stored;

6. The radius r_2 is set to an increased level r_3;

7. Steps 3–6 are repeated, with r_2 replaced by r_3;

8. The stored value $(\dot{Q}_{r_2} C_H) + (r_2 - r_1) C_I$ is compared with
$$(\dot{Q}_{r_3} C_H) + (r_3 - r_1) C_I;$$

9. If the former stored value is less than the latter, it is printed out; if not the radius r_3 is set to an increased level r_4;

10. Steps 7–9 are repeated, etc.

As already mentioned such a process can be classed, only in a very limited sense, as synthesis and the reader may justifiably feel that it is more properly classified as analysis. The following example is certainly more relevant to this sub section.

Size	Department		Association								
3	Arts	*a*	20	4	5	3	7	9	9	2	8
9	Science	*b*	4	20	4	1	6	8	8	2	8
1	Arts/science	*c*	5	4	20	1	6	8	8	2	7
1	Education	*d*	3	1	1	100	5	8	7	2	7
9	Residences	*e*	7	6	6	5	10	8	8	1	8
6	Library	*f*	9	8	8	8	8	100	9	2	8
4	Communal lecture space	*g*	9	8	8	7	8	9	20	1	9
1	Administration	*h*	2	2	2	2	1	2	1	100	6
4	Community facilities	*i*	8	8	7	7	8	8	9	6	100
			a	*b*	*c*	*d*	*e*	*f*	*g*	*h*	*i*

Fig. 4.6. Departmental association matrix

Built-form layout. In the problem of university campus planning described by Drummond, Paterson, and Willoughby (7), two main criteria were involved: *a*, the association between departments of the university and *b*, the characteristics of the site. Fig. 4.6 expresses the association between each department as a numerical value between 0 and 10 and the values in the diagonal elements of the matrix represent the 'internal association' (i.e. the required degree of compactness) within each department; also recorded in Fig. 4.6 is the size of each department expressed as the number of 1000m² plan area modules. Fig. 4.7 gives numerical evaluation of the site; a low value attached to a 1000m² cell indicates that the foundation conditions are

A	B	C	D	E	F	G	H	I	J	K	L	M	N	O	P	Q	R	S	T	U	V	
0	7	7	7	7	0	0	0	0	0	0	6	6	6	0	0	0	0	0	0	0	0	1
0	7	7	7	7	6	0	0	0	0	7	7	7	7	7	5	0	0	0	0	0	8	2
0	6	7	7	7	7	7	7	7	7	7	7	6	7	7	7	6	0	0	0	0	8	3
0	6	6	6	7	6	7	7	7	6	6	3	4	7	6	6	5	0	0	0	8	8	4
0	8	7	6	6	6	6	7	7	6	6	3	3	4	8	7	6	8	8	8	8	8	5
0	8	8	8	8	8	3	5	7	7	7	6	5	0	0	0	7	8	8	0	0	0	6
0	0	0	0	5	8	8	8	8	8	8	7	2	0	0	0	4	8	7	8	0	0	7
0	0	0	0	0	0	0	4	4	4	4	0	2	8	8	2	2	9	8	7	7	7	8
0	0	0	0	0	0	0	2	2	2	0	0	2	8	7	2	2	9	9	8	8	8	9
0	0	0	0	0	0	0	2	2	2	0	0	7	7	7	6	2	2	9	8	8	8	10
0	0	0	0	0	0	0	8	8	4	0	0	7	8	8	7	2	2	8	8	7	7	11
0	0	0	0	0	0	8	9	9	7	8	7	7	8	8	8	2	2	8	8	8	8	12
0	0	0	0	9	9	9	9	8	7	7	7	8	8	8	8	2	2	9	8	8	8	13
0	8	10	10	10	10	8	9	7	8	8	7	8	7	9	6	0	0	0	6	7	8	14
0	5	10	10	10	10	10	10	10	9	9	8	7	8	4	0	0	0	0	6	7	8	15
0	6	7	8	10	10	10	7	6	0	8	7	3	0	0	1	1	0	0	0	8	8	16
0	6	10	10	10	10	10	7	6	0	0	2	2	2	0	0	0	0	0	0	8	7	17
0	6	9	9	9	9	10	8	8	10	10	10	10	10	9	9	4	0	0	5	6	6	18
0	0	8	8	8	9	9	9	9	9	9	9	9	9	9	9	9	9	5	5	4	5	19
0	0	0	8	8	8	8	8	7	9	9	9	9	9	9	9	9	9	4	4	4	5	20
0	0	0	0	0	0	8	8	8	7	7	8	8	8	9	7	0	8	6	6	4	4	21

Fig. 4.7. Foundation conditions of the site (shaded area represents a lake)

unsuitable for building, a high value indicates good foundation conditions.

When it is considered that the possible number of ways of locating 10 building modules on a site which measures 20 modules by 20 modules is 30×10^{18}, some idea of the magnitude of the planning problem is obtained. Even with massive computing power it is impossible to consider all alternative solutions. The technique which must be adopted in the program, therefore, is one of conducting a limited search of the alternatives such that the chance of considering the particular solution which best satisfies the criteria is maximized; this is called a 'heuristic' process.

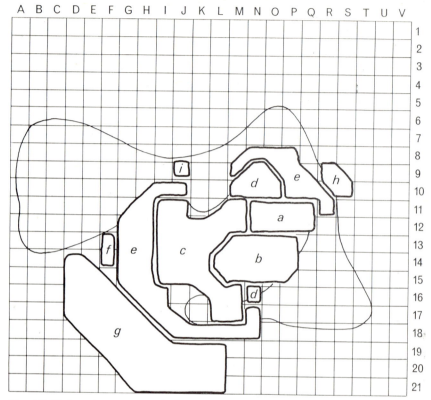

Fig. 4.8. University campus solution

In the university campus problem the first heuristic (or search rule) which had to be formulated was the order in which the departments were to be located on the site. The heuristic adopted was as follows: the department to be placed first is that having the highest association with all other departments; the second department to be placed is that having the highest association with that department already placed; the third department to be placed is that having the highest association with the two already placed, etc. When this order of placing has been sorted out by the computer, the designer locates the modules of the first department on the site grid intuitively (cells G15, G16, H15, H16 on Fig. 4.7). The program proceeds to locate the modules of the second department and of the other departments in turn. The heuristics involved are fairly complex and the interested reader should obtain a copy of the original paper; in principle, however, the location of a module of any department is achieved by a search of the cells in close proximity of those already occupied, for that cell in which the sum of the site value and the association value, modified by a distance weighting, is maximized. Fig. 4.8 gives a graphical representation of the computer output.

Clearly the solution generated will depend on the heuristics adopted, and since there is often no way of telling how appropriate the heuristics are, it is impossible to say how close (or far) the solution is from a true optimum. In the example given, two variables were involved; as the number of variables increases, the heuristics become more dubious until a stage is reached when the proximity of the solution to an optimum may be less than is achievable by purely intuitive methods.

4.3.3. Appraisal

As defined, appraisal implies that a hypothetical design solution exists and the process is one of measuring its characteristics and comparing them with existing criteria. In building design the criteria can be of any one of three types: an optimum, a mandatory limit (for example, 2 per cent daylight factor), or the norm for buildings of similar type. Two examples will be used to illustrate the process of appraisal, one relating to built form layout, the other to the design of a heating network.

Heating network. Consider the problem of sizing a high-pressure hot-water heating system (8) where the location and the output of the radiators has been decided. Initially, the designer hypothesizes a layout for the network. The first task is to compute the required water-flow rates in each segment of this network. A set of equations can be formulated, taking account of the output of the radiators and assumptions regarding the temperature drop over the circuit as a whole and the heat losses from each segment of pipework, which, by a process of iterative approximation, produce the flow rates in each segment. Knowing the water-flow rates and assuming a total pressure drop over the whole network and a distribution of pressure drops round the longest circuit, the diameters of each segment can be calculated.

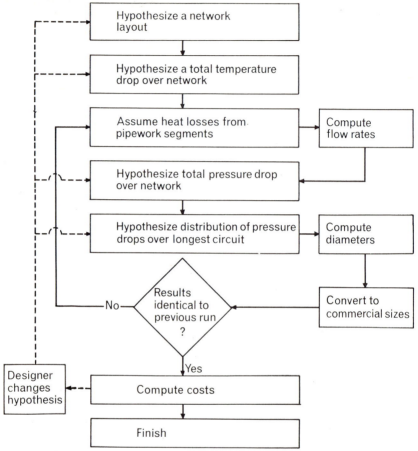

Fig. 4.9. Flow chart for heating system program

When these diameters have been adjusted to the nearest commercially available size, the assumption regarding the heat loss from pipe segments can be improved and the computation repeated. When the outcome remains the same as two successive runs, the solution is costed.

Fig. 4.9 gives the flow chart for a computer program which will perform the sequence of operations just described. It should be noted that the appraisal measure is in terms of cost and that the solution hypotheses have in fact been the layout configuration together with

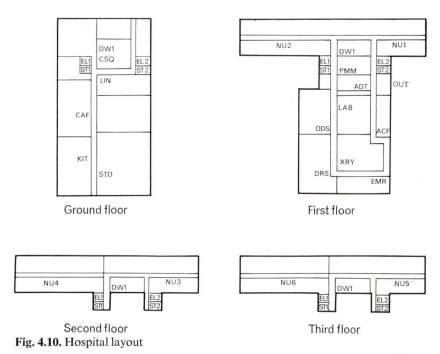

Ground floor

First floor

Second floor

Third floor

Fig. 4.10. Hospital layout

the assumptions regarding total temperature drop over the network, total pressure drop over the network and distribution of pressure drop over the largest circuit.

The broken lines in Fig. 4.9 show how the designer, in the light of the cost appraisal, may return to his design hypotheses and modify them. If no criteria exists for the cost of heating networks, the designer may care to use the program a large number of times to 'generate' comparison data, and to choose, as his design solution, that set of hypotheses which gives minimum cost.

Built-form layout. The example taken to illustrate the process of appraisal in built-form layout is from the work of Souder *et al.* (9). The problem was that of determining the layout of a hospital building based on the pattern of movement of staff, patients, and visitors. The input to the computer consisted of an analysis of data on staff, patient, and visitor movements during each hour of the day, gathered from existing hospitals of similar type, together with a hypothesized layout

P.C.S.—9

giving the location of all rooms, departments, corridors, stairs, and lifts, on each floor. Fig. 4.10 gives some idea of the detail of the hypothesized solution fed into the computer. The computer program was written such that the machine simulated the movement of staff, patients, and visitors round the hospital by tracing the journeys within each hour from source to destination.

The output could be in terms of total time spent in journeys of staff, patients, and visitors within any time period; or use of vertical transport and corridors within any time period; or any other measures which gave an insight into the performance of the hypothesized solution. Thus, if it was found that a particular lift was under-used or that a great deal of the movement was caused by the inappropriate siting of a particular room, the hypothesized solution could be modified and the simulation run again.

In effect the computer is allowing a very fast iterative cycling between synthesis and appraisal which allows the designer to converge rapidly on an optimum solution. In the case of this particular study, the development team had access to a computer with a graphical input device which greatly facilitated the initial input and subsequent modification of the plan layout. If a graphics terminal is not available, however, the layout can be input in terms of co-ordinate points, which, although more laborious, is equally effective.

4.3.4. Summary of programs for design decision-making

Examples have been described of programs used in the decision-making process under the categories of analysis, synthesis, and appraisal. A more complete classification is possible by identifying the stage in design management at which the programs are applicable.

Table 4.3. *Classification of program examples*

	Analysis	*Synthesis*	*Appraisal*
Outline proposals	LP Cluster analysis	Built-form layout (campus)	
Scheme design			Built-form layout (hospital) Heating network
Detail design		Pipe insulation	

Table 4.3 locates the given examples within the framework of the model of the design activity previously proposed.

Appendix 4.1 provides, for the reader who wishes further examples of programs for design decision-making, a list of references under the three headings analysis, synthesis, and appraisal.

4.4. MANAGEMENT PROCESSES

The aim of a management program should be to help promote an environment within which design decision-making can take place. The following subsections deal with certain design office management activities to which the computer is particularly applicable.

4.4.1. Information retrieval

A number of information retrieval systems for the building industry have been proposed and are being developed. The example which will be taken is that developed by Whitton (10) as a working model from which a full-scale information retrieval system could be built. The main characteristics of the system are summarized as follows:

a. It is designed for a remote-access time-sharing computer system;
b. The information is categorized by cost and performance attributes;
c. The input and output is in conversational mode;
d. Storage of information is hierarchial according to the CI/SfB code;
e. The operator can specify cost and/or performance constraints.

The logic of the working model is set out in Fig. 4.11. Since it is only a model, no data yet exists in relation to many of the options; the data which does exist concerns internal partitions and ceiling specifications and Cape Universal ceiling tiles.

To use the system, the designer dials the computer's telephone number and holds a 'conversation' with the machine similar to that outlined in Table 4.4. Initially the computer inquires as to the general area of search—building elements or construction form/materials—and the designer follows the question-mark with his reply. If he does not know the CI/SfB code the machine offers him a multiple choice and then invites him to impose constraints. If the designer declines

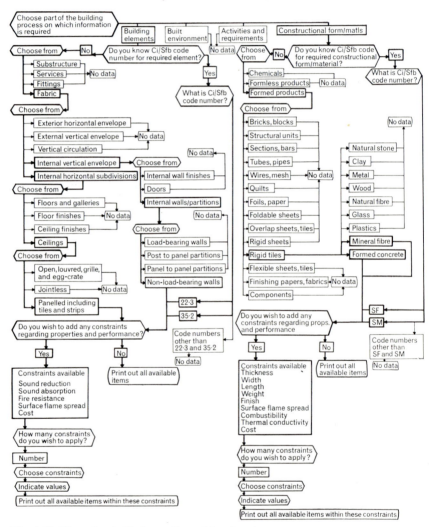

Fig. 4.11. Flow chart of information retrieval system

Table 4.4. *Information retrieval dialogue*

```
ARE YOU LOOKING FOR INFORMATION ABOUT
(1)BUILDING ELEMENTS
(2)CONSTRUCTION FORM/MATERIALS
? 2

DO YOU KNOW YOUR CI/SFB CODE        YES/NO
? NO

FROM THE FOLLOWING LIST CHOOSE ONE ITEM ONLY
(1)FORMED PRODUCTS
(2)FORMLESS PRODUCTS
(3)CHEMICALS
? 1
```

```
CHOOSE FROM
(F)BRICKS,BLOCKS (PER 100)          (E)NATURAL STONE
(G)STRUCTURAL UNITS (NO.)           (F)FORMED CONCRETE
(H)SECTIONS,BARS (M.)               (G)CLAY
(I)TUBES,PIPES (M.)                 (H)METAL
(J)WIRES,MESH (M.)                  (I)WOOD
(K)QUILTS (SQ.M.)                   (J)NATURAL FIBRE
(L)FOILS,PAPERS (SQ.M.)             (M)MINERAL FIBRE
(M)FOLDABLE SHEETS (SQ.M.)          (N)PLASTICS
(N)OVERLAP SHEETS (SQ.M.)           (O)GLASS
(R)RIGID SHEETS (SQ.M.)
(S)RIGID TILES (SQ.M)
(T)FLEXIBLE SHEETS,TILES (SQ.M.)
(U)FINISHING PAPERS,FABRICS (SQ.M.)
(X)COMPONENTS (NO.)
(Y)PRODUCTS IN GENERAL
```

```
EXAMPLE:='KN' MEANS QUILTS,PLASTICS
? SF

DO YOU WISH TO ADD CONSTRAINTS       YES/NO
? NO
```

CI/ SFB	DIMENSIONS THCK	WDTH	LNGH	WT	FNSH	DECOR	FLME SPRD	NON-COMB	K	COST	REF NO
	MM	MM	MM	KG/ UNIT			CLASS		W/M-DEG.C	$/ SQ.M	
SF	6.4	610	610	1.71	PLN	NO	1	YES	.108	.853	3
SF	6.4	610	813	2.23	PLN	NO	1	YES	.108	.807	3
SF	6.4	610	914	2.56	PLN	NO	1	YES	.108	.807	3
SF	6.4	610	1016	2.85	PLN	NO	1	YES	.108	.807	3
SF	6.4	610	1219	3.42	PLN	NO	1	YES	.108	.807	3

Table 4.5. *Information retrieval dialogue (reduced form)*

```
DO YOU KNOW YOUR SFB CODE    YES/NO
? YES

WHAT IS YOUR CODE NUMBER
? 35.2

DO YOU WISH TO ADD CONSTRAINTS    YES/NO
? YES

CHOICE OF CONSTRAINTS
(1)SOUND ABSORPTION (N.R.C.)
(2)SOUND REDUCTION (D.B.)
(3)FIRE RESISTANCE (HOURS)
(4)FLAME SPREAD (CLASS)
(5)COST (£ PER SQ.M. UNIT)

HOW MANY DO YOU WISH TO APPLY
? 1

CONSTRAINT NO.
? 1

CHOOSE FROM (1)VALUE>  (2)VALUE<  AND GIVE VALUE
EXAMPLE:= 1,10 MEANS VALUE>10
? 1,2

THERE ARE NO ITEMS AVAILABLE WITH THIS CODE NO.
AND WITHIN THESE CONSTRAINTS
```

to impose constraints the machine will print out the cost and performance characteristics of all available products together with a reference number; if the computer is asked to type the information under this reference number, further details will be printed out, including the name of the manufacturer of the product. It is important to note that the only responses made by the designer are those immediately following the question-marks.

The 'conversation' with the computer will be rather briefer, of course, if the designer knows the CI/SfB code. The output will also be less if cost and/or performance constraints are imposed. Table 4.5 records the designer/computer exchange for a situation in which the CI/SfB code is known and the constraints set are too tight.

The potential for such a system is considerable. If one imagines the product information service as equivalent to the standard library programs for which the computer bureaux are at present responsible, the equivalent of the designer's own programs could be a cost- and performance-in-use file built up and constantly added to from experience in particular cases. Thus all subscribers could have access to the product information files while each subscriber could have access only to his own user experience file.

Whitton describes information as 'a highly perishable commodity' which must be available when required in the correct form at the right price. The advantages of the system described above, over the conventional bulky, ill-assorted, incomplete, often out-dated catalogue system which is operable by only one member of the office staff, are obvious, and will increase as the rate of information production increases.

4.2. Critical path analysis (CPA)

It is only within the last few years that architects have realized that the complexity of design management and of construction management warranted the application of reasonably sophisticated management techniques. One technique which is now used extensively is that of Critical Path Analysis (CPA); like the similar technique, Project Evaluation and Review Technique (PERT), CPA is concerned with finding, in a network expressing the tasks to be undertaken and their interdependency, the longest path through the network since this represents the shortest time within which the project can be completed.

For illustration of the principle, a simple exercise concerning the management of the construction of a village hall will be given. As a starting-point the event which marks the completion of each task to be undertaken is given a number. Next, the dependancy of any event on any other events is recorded as follows:

Table 4.6.

Event number	Description	Dependent on events
1	Start	—
2	Delivery of frame completed	1
3	Foundations completed	1
4	Erection of frame completed	1, 3, 2
etc		

This record allows the basic network to be drawn as in Fig. 4.12, in which the squares represent events and the links represent activities. At this point the only figures on the network are those in the top quadrant of each square, viz. the event numbers. The next step is to attach to each activity an estimated duration. When this has been

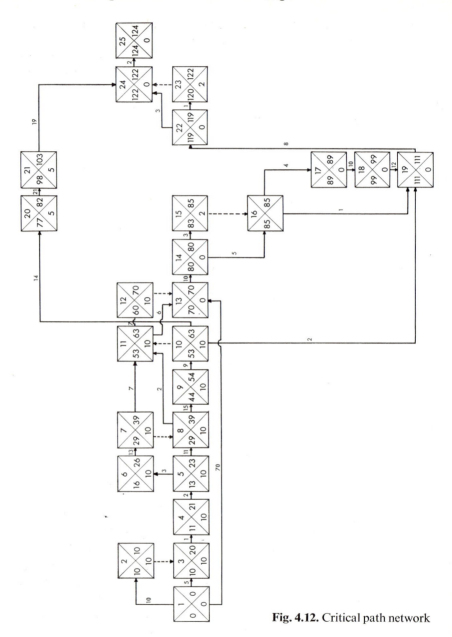

Fig. 4.12. Critical path network

done it is possible to enter in the left-hand quadrant of each square starting at event 1, the earliest time at which each event can be completed. This gives a value of 124 in the left-hand quadrant of the square representing event number 25. The next step is to work back from event 25, subtracting the activity duration values and entering, what is in fact, the latest finishing time, in the right-hand quadrant. The bottom quadrant can now be completed by entering the difference between the earliest and latest finishing time, which is called the 'float'.

The 'critical path' is that route joining the events for which the float is zero, in this case 1, 13, 14, 15, 16, 17, 18, 19, 22, 24, 25. The broken lines in the network diagram are called 'dummy' activities which take zero time but which represent an interdependency between two events.

The procedure which has just been described can of course be done manually. When a large complex project is under analysis, however, the labour involved is considerable; the real advantage offered by the computer is the up-dating of the network if, in practice, any particular activity takes longer than anticipated due to weather or late delivery of materials or illness in the work-force. Standard library programs exist which accept input data as in Table 4.6 together with activity duration times and output the critical path and the 'next-to-critical' paths.

4.4.3. Specification writing

The amount of labour taken up in writing the specifications which accompany design drawings can be considerable; while the specifications vary from one job to another, the general format, and indeed certain details, remain approximately the same. Harper (11) has developed a computer system which automates, to a large extent, the tedious chore of specification writing. The principle is one of producing a master file of specification sections with each line identified by a code number. To prepare a set of specifications for a particular job, the designer encircles on a print-out of the master file the code numbers of the lines which are to be deleted (i.e. he deletes those alternative forms of specification which are inapplicable to the particular job). These code numbers are typed up by the computer operator and fed into the machine. The program which controls the production of the particular specification prints out the amended master copy (with adjustments made for paragraph numbering or lettering) on to a stencil or off-set master which may then be used for duplicating by standard reproduction techniques.

If for a particular case it is desired to add a word or section, the appropriate addition is typed into the computer together with a line code number which ensures its interpolation in the correct position. At the same time the master file is altered in similar manner to ensure this particular option is available for subsequent projects.

4.4.4. Bills of quantities

The production of bills of quantities by computer has been studied by a number of private firms (12), by Local Authority Consortia, and by the MPBW; the system proposed by MPBW is published in a research and development paper (13).

Whatever the system, it starts with a decision as to a coding of elements, components, finishes, etc., which will be used on every job. As the quantity 'taking-off' proceeds, the codings are added. Dimensions and codes are now punched up on cards or paper tape and verified. The processing of the data may include the arithmetic tasks of squaring and cubing; sorting into code order and within code by size; adding together like items; checking for errors in the units of movement; etc. The output can then be printed on to stencils or offset lithography masters for duplication and sending out for tendering. Alternative collations for the purposes of costing analyses, operational analyses, or CPA scheduling are possible as is the preparation of the final account.

4.4.5. Production drawings

The availability of plotting machines, which are driven and controlled by computers, offers the designer an extremely efficient way of producing working drawings of his scheme. Plotters are of two types: the flat-bed plotter in which the arm actuating the pen moves in two dimensions; and drum plotters in which the pen arm moves along the longitudinal axis on the surface of a drum and the drum, round which the paper is wrapped, rotates to give movement in the second dimension. As Fig. 4.13 shows, the accuracy and clarity of the drawing and lettering is more than adequate for production drawing purposes.

Assuming that the architect produces, at some stage in the design activity, rough plan and elevation sketch drawings of the proposed

Fig. 4.13. Graph plotter output

scheme, there are two basic ways these sketches can be fed into the computer to furnish data for a production drawing program.

The first method involves punching up the two-dimensional co-ordinates, relating to an origin, of points on the sketch at which lines start or finish; these co-ordinate points are code numbered sequentially. This information is followed by a list of the lines which are to be drawn, each being specified by the code numbers of the points at both ends. The third item of data is the required lettering, together with the co-ordinates of the first letter and the axis along which it is to be printed.

The alternative method is to use a 'pencil-follower' or 'trace-reader'. This piece of apparatus produces coded data on perforated paper tape when a cursor is moved to any position on the sketch and a button pressed. The facility exists for interpolating data between the punching of any two points to allow a statement of the fact that there is a line joining the two points and to permit a coding of its characteristics.

The program can be arranged to regularize the co-ordinate data to a prespecified module. This allows data lifted from a crude sketch to be 'squared-up' and made accurate. Standard programs are available for most makes of machine which will take the data and produce, on the plotter, an accurate diagram to any scale within the limits of the paper size. One of the facilities of the computer/plotter system is that of selected information on the output drawing; all service runs can be shown within the plan layout, or only selected ones; any service distribution can be drawn independent of the plan layout, etc.

By extension of the system described above, any selection of perspective drawings can be produced on the plotter. Again standard programs exist and the only additional data required is the third co-ordinate of each point, the co-ordinates of the viewed point and the co-ordinates of the viewing point.

In those design offices which are lucky enough to have a computer with a cathode ray tube graphics terminal, the process is even simpler, since design solutions are sketched directly on to the surface of the tube. The computer thus knows the co-ordinate points and can produce drawings automatically on the plotter.

Appendix 4.2 provides a list of references to management process applications for those who wish further reading.

4.5. FUTURE DEVELOPMENTS

So far a fair number of disparate problem areas have been reviewed within both the decision-making process and the management process. The application of computers to these problem areas has been piecemeal and one would be hard put to it to find any two programs written in the same language for the same machine. The development of more and more sophisticated programs for certain problem areas is proceeding rapidly while other problem areas are entirely neglected. The point would appear to have been reached when, unless a major effort is made to co-ordinate development, taking full cognizance of what the future holds, both in hardware development and in changes in professional working, the promise offered by computers for an improved level of built environment will not be realized.

4.5.1. Government policy

While the products of the building industry represent 50 per cent of the capital items of the Gross National Product (which is equivalent to 16 per cent of the GNP as a whole), the industry is sadly segregated in terms of professional responsibilities and attitudes; even within a single profession, such as architecture, effort and direction is dispersed over a large number of small firms. In a situation such as this the chances of a co-ordinated approach are slim unless government agencies are prepared to play a decisive role. But even at government level divisions of responsibility hamper action: while the MPBW has a responsibility to the building industry, another ministry, the

Ministry of Technology, has prime responsibility for computer hardware and applications.

MPBW has a standing committee on the application of computers in the construction industry (CACCI) (14–22) which operates under the following terms of reference:

to keep under review the present and potential application of computers in the construction industry, taking account of experience and trends in other industries; to identify areas in which there is a need for co-ordination of effort; to examine the need for action to secure the wider application of existing knowledge and the promotion of further advances; and to make recommendations.

A number of subcommittees have been formed—on data coding, on structure engineering, on quantity surveying, on services engineering, and, more recently on architectural applications. The report (23) of three working groups on the application of the computer to architecture, set up by CACCI, was published recently.

Apart from a grant to develop a co-ordinated structural engineering computer system (GENESYS) (24) and a few awards for isolated architectural applications, nothing of major import to architecture has emanated from MPBW.

4.5.2. Changes in professional practice

Many schools of architecture are implementing a two-degree curricular structure. The products of these schools should be better versed in the techniques of scientific investigation and in the analytical processes necessary to study the building as a complex interactive system. They will (hopefully) be less inflexible with regard to disciplinary demarcation and more able to see the problems of the industry in general terms.

It is predicted that the growth in the number of 'package dealers' and in the number of management consultancy firms showing an interest in building design problems, will bring about the differentation, which exists in other design disciplines, between specification designers and product designers.

As 'repeat' clients (local authorities, large corporations, etc.) take a greater proportion of the industry's effort, the pressure will grow to design to performance specification. There will be an increased awareness of the need for continual redesign (or design-in-use) throughout the life of a building to meet the changing requirements of the client organization. The greatest challenge of all, however, will come from the growing demands of those affected by environmental design decisions to play some part in the making of the decisions, i.e., a demand for an effective participatory planning mechanism.

4.5.3. Integrated systems

A number of integrated computer-aided design systems have been proposed and a few have reached a modest state of development (25, 26). The system which will be discussed here is that under development by the Architecture and Building Aids Computer Unit, Strathclyde (ABACUS) (27) which goes some way towards relating to the models of the design activity described at the start of this chapter and towards the challenges of the building design profession of the future.

The basic tenets of the ABACUS approach are: that the synthesis, or generation of design solutions is best left to man and that the computer should be applied wholeheartedly to the task of rapid and explicit appraisal; that the crux of the design optimization problem lies in the balance struck between the investment costs and *all* the performance variables; and that the client/user organization, provided with the explicit appraisals of alternative solutions, is in the best position to exercise value judgement.

Three program packages will be developed, supported by an information file. The first will be relevant to the 'outline proposals' stage in management, the second to the 'scheme design' stage, the third to the 'detail design' stage. The following subsections deal with the elements of the first package.

Input. The input consists of:

I1. A site analysis, similar to that shown in Fig. 4.7 but incorporating ordinance heights;
I2. The geographical location of the site (chosen from a list offered by the computer);
I3. The building type (drawn from a list offered by the computer);
I4. The number of occupants;
I5. A list of the departments in the scheme and a matrix of the association between them;

I6. The geometrical description of the proposed scheme in three dimensions (each department of the scheme entered separately by defining the co-ordinates of the opposing vertices of the 'block');

I7. The constructional system (chosen from a list offered by the computer);

I8. The construction materials (chosen from a list offered by the computer);

I9. The over-all percentage area of glazing;

I10. The orientation of the scheme to the north point;

I11. Proposed building life.

Filed information

F1. Indoor–outdoor monthly temperature differences for selected geographical locations and altitudes;

F2. Monthly solar radiation per hour for selected geographical locations;

F3. Hourly vector co-ordinates of the sun for one day in each month for selected geographical locations;

F4. Maximum storm water density for selected locations;

F5. Capital costs of selected constructional systems and materials as a function of unit floor area and wall area;

F6. Cost per unit of energy generated from different fuels;

F7. Cost limits for selected building types as a function of occupancy or floor area;

F8. Cost per cubic unit for cut and fill; for dumping and for purchase of fill;

F9. Maintenance costs as a function of unit wall area and floor area;

F10. Thermal transmittance values for glass and selected materials;

F11. Water consumption per person per day for selected building types;

F12. Lift numbers as a function of occupancy per floor;

F13. A set of initially empty files, the purpose of which will be described later.

Program constants

C1. Distance from perimeter for PSALI;

C2. Distance from perimeter for artificial ventilation;
C4. Area of boiler house floor per unit of boiler power.

Output

O1. Capital cost of the scheme in present worth and in annual payments over its life; check against cost limit;
O2. Running costs categorized as maintenance, heating, cooling, and lighting in present worth or in annual payments over the life of the scheme;
O3. Costs of cut and fill in present worth on annual payments;
O4. Monthly heat gain and loss over the year;
O5. Required water storage;
O6. Number of lifts in each lift bank;
O7. Maximum roof drainage rate;
O8. Site utilization;
O9. Plot ratio;
O10. Compactness (defined as the ratio of the surface area of the scheme to the surface area of a hemisphere of equal volume);
O11. Hours of sunshine on each surface in each month of the year;
O12. Best choice of fuel;
O13. Boiler-house size;
O14. Sum of the quotients of association and distance of each 'department'.

The format of the other two packages is analogous but the detail of input and output will of course be greater and will encompass internal traffic, location of service outlets and distribution, structure, etc. The main difference will lie in the format of the geometrical description of the scheme: in the first package, 'blocks' of space are described; in the other two packages the surfaces which bound the spaces will be described to allow specification of the properties of external and internal wall components.

The mode of operating the packages is as follows:

1. The architect hypothesizes a design solution.

2. This is fed into the computer which outputs the appraisal measures.

3. The architect and the client/user organization consider the measures; initially there will be little normalized data to compare with the measures. The purpose of the empty files, however, is to allow the

measures obtained for each scheme to be retained as data for future reference. Thus there is a two-way exchange between the program and files.

4. On the basis of the evaluation of the initial scheme by the architect and the client/user organization, the architect modifies the scheme and has it reappraised by the machine.

5. This cyclic process continues until the architect and client/user organization are satisfied with the balance between the cost investment and the performance return.

6. The co-ordinate description of this final scheme is used as the input for a program to output production drawings on the graph plotter.

7. The co-ordinate description of the final scheme at the detail design stage is used for 'taking-off' of quantities and the production of the bills of quantity.

The first of these three packages is in working order and is being commissioned by a working group of practitioners. It is hoped that when remote graphics terminals become economically viable, the system can be converted from the present keyboard terminal system.

Apart from using the packages for the design of specific buildings, they can be used as an analytical tool to study response of client/user groups to alternative built environments. The architect generates, say, seven different schemes and presents the computer appraisals of the scheme to the client/user group which comprises, say, eight members. Each member of the group is given seven votes to cast as he wishes for any of the alternative schemes. Consider that Table 4.7 represents the

Table 4.7. *Result of client/user group voting*

	Group member								
Scheme	A	B	C	D	E	F	G	H	Total
1	0	2	6	0	1	0	0	0	9
2	7	1	0	1	1	0	0	0	10
3	0	2	0	1	1	0	5	1	10
4	0	1	0	2	1	3	2	0	9
5	0	0	1	2	1	0	0	4	8
6	0	0	0	1	1	4	0	0	6
7	0	1	0	0	1	0	0	2	4
Total	7	7	7	7	7	7	7	7	56

voting. The procedure then is to feed back to the group the results of the voting and invite them to re-vote. This new result is fed back and the process continued until two consecutive 'runs' produce the same result, i.e. until each member is as satisfied as he possibly can be with the results.

This procedure results in a polarization of preferences and the manner in which individual group members modify their voting gives an insight into the 'trade-offs' he is prepared to make between alternative schemes.

Yet another use of the packages is as a means of generating a large volume of data on cost and performance measures. With a sufficient volume of data there is a chance of establishing the causal relationships between cost and performance. If such relationships are eventually established, the main obstacle to logical building design will have been overcome.

ACKNOWLEDGEMENTS

The author wishes to thank the editors of *Building* for permission to reproduce Fig. 4.11 and Tables 4.4 and 4.5 and California Computer Products, Inc, for permission to reproduce Fig. 4.13. Acknowledgement is made of the inclusion, in Appendices 4.1 and 4.2, of extracts from a *Bibliography on the application of computers in the construction industry 1962–1967*, published by the Ministry of Public Building and Works. Finally the author wishes to thank Professor Thomas A. Markus and the members of the Architecture and Building Aids Computer Unit, Strathclyde (ABACUS) for help and encouragement.

REFERENCES

1. MARKUS, T. A. (1967). 'The role of building performance, measurement and appraisal in design method', *Architects' J.* **146**, 25.
2. —— Ibid.
3. ROYAL INSTITUTE OF BRITISH ARCHITECTS (1967). *Handbook of Architectural Practice and Management* (London: RIBA Publications).
4. SINGLETON, W. T. (1966). 'Current trends towards systems design', *Ministry of Technology: Ergonomics for industry*, 12 (London: HMSO).

5. KREJCIRIK, M., and SIPLER, V. (1965). 'Use of computers for determining the optimum development pattern for a residential area', *BRS Library Communication*, no. 1235 (May).
6. MILNE, M. (1968). 'CLUSTR: A structure finding algorithm', Paper presented for the *Design Methods Group Conference, Cambridge, Massachusetts* (June).
7. WILLOUGHBY, T. *et al* (1970). 'Computer aided architectural planning', *Operational Research Q.* **21,** 1.
8. HOWARD, H. (1957). *On pipe sizing accelerated hot-water systems with an electronic digital computer*, Associateship thesis, National College of Heating, Ventilation, Refrigeration and Fan Engineering (August).
9. SOUDER, J. J., *et al.* (1964). *Planning for hospitals: a systems approach using computer aided techniques* (Chicago: American Hospitals Association).
10. WHITTON, D. (1969). 'A working model of a computer-based information retrieval system', *Building*, **ccxvii**, 6605.
11. HARPER, G. N. (1966). 'Autospec: automated preparation of specifications', *Jl of Structural Division, American Society of Civil Engineers* (December).
12. MONK, K. W., and DUNSTONE, P. H. (1968). 'Quantity surveying by computer 1961–1968', *Building*, **ccxiv** (6524), 67–8.
13. FARRAR, C. H., and MALTHOUSE, R. F. W. (1969). *The preparation of bills of quantities with the aid of computers* (London: Ministry of Public Building and Works).
14. MINISTRY OF PUBLIC BUILDINGS AND WORKS (1968). *Computers in the construction industry* (June) (London).
15. —— (1968). *Construction, education and the computer* (Sept.) (London).
16. —— *A study of coding and data co-ordination for the construction industry* (London: HMSO).
17. —— *Coding and data co-ordination: a short report* (London: HMSO).
18. —— (1967). *The application of computers in the construction industry—a review of the present position 1967* (London).
19. —— (1967). *Computers in construction: conference report, London* (April).
20. —— (1967). *Computers in construction: conference report, Manchester* (Oct.).
21. —— (1968). *Computers in system building: conference report, Nottingham* (Jan.).
22. —— (1968). *Bibliography on the application of computers in the construction industry 1962–1967* (London: HMSO).
23. —— (1969). *Computer-aided architectural design, parts 1 and 2* (London).
24. —— (1968). *GENESYS; conference report, London* (Nov.).
25. ROOS, D. (1966). *ICES system design* (MIT Press).
26. SOMMERFELD, W. F. (1967). *A graphical information system for use in building design* (June) (Department of Civil Engineering, MIT).
27. MAVER, T. W., and MCKENZIE, B. (1970). PACE 1: an on-line design facility. ABACUS *Occasional Paper* no. 4. School of Architecture, University of Strathclyde.

Appendix 4.1

SELECTED BIBLIOGRAPHY OF DECISION-MAKING PROGRAMS

Analysis

ALMENDINGER, V. V. (1964). *SPAN: a system for urban data management*, AD-611078, 25 pp.
Describes SPAN (Statistical Processing and Analysis) system, programmed for the IBM 7090/7094. Its development was initiated at the Penn-Jersey Transportation Study to support a comprehensive simulation model of regional growth.

ALEXANDER, C. (1964). *Notes on the synthesis of form* (Harvard University Press).
Describes the application of a computer to solve the hill-climbing process used in decomposition of a design problem's parameters into subsets characterized by a minimum of interaction between subsets; programmed for an IBM 7090.

MOSELEY, L. (1963). 'A rational design theory for planning buildings based on the analysis and solution of circulation problems', *Architects' J.* **138**, 525–37, 4 refs.
Article describes in detail how linear programming may be used by the architect to solve complex problems of circulation. Part 1 gives terms in which problem must be stated to meet requirements of the technique; then the concepts behind its application are described, and the technique itself. Finally, complete design procedure is summarized as a sequence of steps leading to an architectural solution which can be assessed in terms of its circulation efficiency.

THOMSEN, C. (April 1965). 'How high to rise', *Am. Inst. Architects' J.*, 66–8.
Case-history of an analysis to determine the optimum economic height of a high-rise office building, performed on a computer. It was found that the computer increased design freedom, that it provided information, and that it is an important tool, but not substitution for, the architect.

Synthesis

MOORE, J. M., and MARINER, M. R. (1963). 'Layout planning: new role for computers', *Mod. Mater. Handling*, **18**, 38–42.
Gives general account of a digital computer technique for deciding the location of a new facility to be added to an existing complex with the object of minimizing movement costs between the new facility and those it serves. The technique can be used for locating a new production unit or service area in a factory layout, a new building, or a new warehouse in a distribution system. One actual application of the technique is described.

SINYAKOV, Y. (Sept. 1965). 'Can a machine create a design?', AD-622387, 7 pp. (unedited rough draft translation from *Pravda*, 5 July 1962).
The use of computers for assistance in making architectural designs is described. Andrev Savchenko, of the Moscow Architectural Institute, is credited with using cybernetics in calculating the strength of structural slabs for roofs and for covering a winter stadium. The machine is said to be able to create design; by 'reading' an architectural sketch into a computer, an image of all possible combinations of architectural forms is attained from which a selection may be made.

WHITEHEAD, B., and HAFEZ, E. (Jan. 1966). 'A systematic method for warehouse design', *PACE*, **1** (1), 13–20.
Describes method of estimating minimum over-all costs for building a warehouse: economic future stock levels were calculated for each commodity and were used as basic data in deciding the volume, shape, and internal arrangement of proposed building. Building and operating costs were calculated for each possible solution and the one with the combined minimum was chosen. The design method has been programmed for a KDF-9 computer.

WRIGHT, J. (1967). 'Automatic design of plane steelwork frame', paper presented at *4th International Congress on Application of Mathematics in Engineering* (Weimar, June), 14 pp.
Paper describes a computer program which automatically designs a plane steelwork frame under multi-load conditions. The approach to automatic design is explained in detail. The basis of the design process is to perform automatically the approach of a designer in normal drawing-office procedure, if the designer has been given unlimited time and reasonable computational power. An illustrative example is given to show the complete computer output.

MOUČKA, J. 'Decision making in the initial phase of design', *Architects' J.* (to appear).
From an association matrix a plane graph is produced. The dual of this plane graph is used to produce a plan layout.

BEAUMONT, M. (1967). *Computer aided techniques for synthesis of layout and form with respect to circulation*, PhD thesis, University of Bristol.
A large set of spatial elements is decomposed by cluster analysis into two subsets which are located, heuristically, on a grid; in turn each of these subsets is decomposed and the subsets located; this process is continued until all the elements of the original set have been placed.

MAVER, T. W., and FLEMING, J. (1970). 'SECS: A programme package for accommodation scheduling in comprehensive schools.' ABACUS *Occasional Paper* no. 2. School of Architecture, University of Strathclyde.
A program suite is described, which produces and costs an optimum schedule of accommodation given a statement of group sizes in each subject.

Appraisal

PARLOW, H. (1966). 'Lift operation and computers: a simulation of performance', *Architects' J.* **143**, 747–53.
Describes an investigation of the performance of a lift installation at Imperial College, London. Computer programs were written for an Elliot 803 in order to simulate various facets of lift operation. These included arrival pattern of people at the lifts, journey pattern, number of floors served, and journey times. Each computer program was used as a simulation model, thus enabling one or more of the parameters to be altered in order to determine the effect which this has on the others.

NEWMAN, W. M. (1966). 'An experimental program for architectural design', *Computer J.* **9**, 21–6, 4 refs.
Describes a program which shows how a computer-driven display could be used as an input/output device when designing buildings with modular industrialized building units. It was written for a PDP-7 equipped with a Type 340 Display and Light-Pen. The user

can add wall units, windows, doors, etc., to his design by using a teleprinter. A picture of the unit then appears on the screen, can be moved about and inserted permanently in the design. The program can perform numerical processing on the design, including calculating areas.

CAMPION, D. (1966). 'Design simulation by computer', *Architl Rev.* **140**, 460–4.
Describes in detail the use of a mathematical model to simulate the operation of a self-service dining-room and its associated servery.

TAYLOR, J. (1967). *The science lecture room: a planning study to examine the principles of location and design of lecture rooms in the development of university science areas*, 119 pp., 84 refs. (Cambridge University Press).
This report suggests methods and principles of planning for lecture-rooms within a university science area. A series of models for groupings of rooms within an over-all development was made and a computer program, 'Seatpoint', set up as a means of comparing arrangements which would satisfy particular design criteria as regards flexibility, size, utilization, etc. Selected print-out of this computer program is presented.

RAWLINGS, B. (1964). 'A computer analysis of structures under impulsive loading', *Int. Ass. Bridge Struct. Engng Prelim. Publ. 7th Congress*, *Rio de aneiro*, pp. 45–54
Paper presents a method of analysis which allows a rigid framed structure to be examined in terms of its deformation under system of time-dependent applied forces. By formulating the equations in matrix form, the method becomes convenient for application to digital computation, as only standard matrix manipulations are involved.

BIGGS, J. M., and LOGCHER, R. D. (May 1964). 'STRESS: a problem-oriented language for structural engineering', AD-604679, 23 pp.
Stress is a general purpose programming system for the analysis of structures. It has three distinguishing characteristics: (1) input language is that of the structural engineer, which makes possible direct communication between the engineer and the machine; (2) system is capable of analysing a wide variety of structural types and loading conditions, thus permitting industrial use on a routine basis; (3) design process is expedited by the fact that modifications of original structure for alternate designs can be easily executed.

ROOS, D. 'An integrated computer system for engineering problem solving', *Jt Computer Conf. Fall 1965 Proc.*, Part 1, 423–33, 13 refs.
Discusses the main features of ICES, the integrated computing system for total civil engineering problem solving, being implemented for the IBM System/360 Model 40 at MIT. ICES is a modular system designed to enable the engineer to communicate and interact with the computer easily. The programmer uses Icetran (extended Fortran) language to develop and modify the necessary components of the system. Dynamic memory allocation, alternate data structures, and data transfer and management facilities are available to the programmer.

SHARMA, S. P., and GOYAL, B. K. (1966). 'Analysis of large pin-jointed space frames', *Indian Concr. J.* **40**, 288–91.
This paper presents an iterative procedure for the analysis of large pin-jointed space frames using deformation distribution. Using this approach, large frames can be analysed even on small computers. The procedure has been programmed on an IBM 1620 model 1 computer. Space frames having a maximum of 400 joints (excluding supports) can be analysed on a 60K machine.

WEAVER, W., and NELSON, M. F. (1966). 'Three-dimensional analysis of tier buildings', *Am. Soc. Civ. Engrs Proc.* **92** (ST6), 385–404.
Three-dimensional analysis of tier buildings using a digital computer is feasible. Errors inherent in the two-dimensional approach may be significant, whereas the three-dimensional analysis contains no theoretical discrepancies. The analytical model of the structure is examined in detail. The stiffness method of solution is described, followed by a brief outline of the computer program. Results for an example 20-storey building are given.

RAO, K. R., and PRAKASH, CHANDRA (1966). 'Digital computer determination of thermal frequency response of building sections', *Bldg Sci.* **1**, 299–307.
Presents a digital computer program in Fortran for the determination of the thermal frequency response characteristics, i.e. amplitude decrement and phase lag of building sections. Matrix method has been employed. Program is quite flexible so that composite sections of any number of layers and frequency response characteristics for any desired number of harmonics can easily be handled. Typical data for a few homogeneous and composite sections obtained by IBM 1620, are presented.

DUNSMORE, B. J. (30 Nov. 1966). 'A computer program for the Hardy Cross analysis of water distribution system', *Wat. Sewage Wks*, **113**, R111–13, R117–18, R120, 6 refs.
Describes in some detail, and gives complete listing of a program written in Automath-800 Honeywell Compiler Fortran II language, for the Hardy Cross analysis of a water-distribution system.

ASKARI, J. S. (Sept. 1967). 'Universal building complex load and energy simulation computer program', *Heat. Pip., Air Cond.* **39**, 106–12.
A general computer program to determine accurately the energy requirements of any air-conditioning system for any building or building complex anywhere in the world has been developed. It is capable of simultaneously processing up to 750 combinations of construction units and surface orientations with ten alternate air-conditioning systems.

BLOCK, B. (Dec. 1967). 'PIFL: computer program for design and analysis of pipe flow systems', *Heat., Pip., Air Cond.* **39**, 71–6.
Discusses the PIFL computer program which has been developed to handle common calculations relating to flow and pressure drop in piping systems. The program was written for the IBM 1620 with 20K memory.

MAVER, T. W., and McKENZIE, B. (1970). 'Simulation of air-terminal and maternity suite by computer', Internal memorandum, School of Architecture, University of Strathclyde.
The application of queueing theory to the sizing of facilities in an air terminal and a hospital maternity suite.

Appendix 4.2

SELECTED BIBLIOGRAPHY OF MANAGEMENT PROGRAMS

ANON. (April 1966). 'Network analysis calculations by computer', *Consult. Engr*, **30**, 88–91.
Describes how calculations associated with network analysis can be economically and readily solved by computer. Factors involved and an example of the way in which a computer was used in conjunction with network analysis, for calculations on the Victoria Line project, are given.

MILLER, P. F., and CORDINER, I. D. (Nov. 1966). 'Presentation of critical path analysis results using the Cascade Activity Numbering (CAN) method', *Civ. Engng Publ. Wks*, **61**, 4 pp.
Detailed description of CAN which is a serial listing of activities producing a stepped diagram in the shape of a waterfall or cascade. An IBM 1440 program has been developed to produce Cascade from precedence diagrams, and authors are engaged on development of a computer construction programming method. The basic method is given.

MINISTRY OF PUBLIC BUILDING AND WORKS (1967). 'Network analysis in construction design', *R and D Building Management Handbook*, no. 3, 40 pp., 9 refs. (HMSO).
Report prepared by Joint Working Group of RIBA, RICS, ICE, and ISE. It describes the technique of network analysis and its application to the programming and control of the design process in the construction industries. It includes a chapter on analysis by computer, with examples of input and print-out.

UNITED NATIONS ECONOMIC COMMISSION FOR EUROPE (1967). 'The use of electronic computing machines in programming organisation and management within the building industry', Committee on Housing, Building and Planning, 5 pp., restricted (Geneva).
Report includes proposal by the UK to consider the application of computers under two main headings: to over-all planning and programming in the construction industry, and to the building process.

BIRGERSON, B. (1967). 'CBC/BDC as a national system', *RIBA J.* **74**, 394–9.
Article by the Vice-President of the Building Data Council in Stockholm. Describes work of the body set up in Sweden to make computer services for contract documentation available throughout the Swedish building industry.

BARNETT, J. (1965). 'Computer-aided design and automated working drawings', *Architl Rec.* **138**, 85–93, 97 refs.
Describes main graphic aids, with particular reference to Sketchpad. The Digigraphic Display Program, DAC, and MAC are also mentioned.

CLARK, W. E. (1966). 'Graphical input and output forms', *Bldg Res.* **3** (2), 27–31, 6 refs.
After stating over-all need for communications relating to the design and construction process in building, author describes various forms of computer graphical input, and some output techniques.

LESLIE, W. H. P. (1967). 'Computer graphical design aids', *R. Aeronaut. Soc. J.* **71**, 237–44, 14 refs. Paper presented at the Symposium on the Impact of Digital Computers on Engineering, London (April 1966).
A detailed illustrated survey of the range of computer graphical design aids now being used and developed.

BOLT, BERANEK & NEWMAN INC. (July 1965). 'Computer-aided checking of design documents for compliance with regulatory codes', PT-174095, 54 pp.
This paper summarizes a preliminary study of the application of computer technology to the function of checking design documents for compliance with regulatory codes.

ALEX GORDON AND PARTNERS (April 1966). 'Job costing and expense analysis by computer', (handout), 1 p.
Brief description of the system of Computerised Analysis and Recording devised by Alex Gordon and Partners, architects. The system is applied to job costing and expense analysis and has been in successful operation since 1965.

ASH, L., *et al.* (4 Nov. 1967). 'Rationalisation in the office', *Nth. Archit.*, 878–88.
The authors describe a system of project management based on network analysis and critical path methods, but using a regular monitoring and updating process by computers to overcome the limitations of basic networks. They describe in some detail, with the help of several diagrams, the entire program process, from the preparation of the network to the content of scheduled reports, and conclude with a summary of PERT cost and 1440 project control systems.

BLACKMAN, T. J. (1967). 'Production of perspectives by computer', *Consult. Engr*, **31**, 57–8.
Description of use of computer graphics in production of perspectives. Computer methods, data preparation, types of display, and shading are discussed.

WINDROSS, R. S. (1967). 'Preparing bills of quantity on a computer', *O & M Bull*, **22** (1), 30–42.
Describes the methods of computer-based bill of quantity preparation and production, and outlines the principles of bill preparation. Coding structure and the handling and processing of data are described and the advantages of the system are considered. Possible methods of reducing the expense and time required on the input coding stage are discussed.

ROYAL INSTITUTION OF CHARTERED SURVEYORS' RESEARCH & INFORMATION GROUP OF QUANTITY SURVEYORS' COMMITTEE (1967). *Computer techniques: Conference July 1967*, 206 pp., bibliography.
Papers cover standardization and the following computer systems for quantity surveying; AUTOBILL; CLASP; Gleeds Link System; Monk and Dunstone; MPBW; ICT Bournemouth CBC; Hertfordshire CC; CEGB Development Group of Chartered Quantity Surveyors.

BRITTEN, J. R. (1967). 'Computerising resource needs', *BRS Current Papers*, Construction Series 39 and 39A (reprinted from *Building*, **ccxii** (6457) (1967), 125–6 and 129–30.
Outlines some resource aggregation scheme requirements, and a computer program to meet them. Program facilitates study of alternative assumptions of project schedule and activity requirements by presenting consequences of any schedule or decision. Expenditure can be compared with estimate. Cumulative cash flow is calculated including expenditure on all resources, and income received. This too can be compared with cash flow. Program can be used to monitor individual projects, or a company's work-load.

WATES, C. (April 1967). 'Computers for contracting', paper presented at the *Conference on Computers in the Construction Industry*, sponsored by the Ministry of Public Building and Works, 20 pp. (London).
A survey of the fields in which computers can be valuably used by contractors to control building operations in the broadest sense, illustrated by examples from Wates's own experience. Two further experimental applications are described: to the plant yard, and to an overhead reporting system.

5

The weathering of organic building materials

H. E. Ashton
Division of Building Research,
National Research Council of Canada, Ottawa

5.1. INTRODUCTION

The ability of materials used in the construction of buildings to resist changes caused by the action of the weather is of great interest to the architect, the builder, and the ultimate user. This article will consider the effects of the elements of the weather on those building materials classed as organic.

Although there are ninety-two natural chemical elements, relatively few are plentiful enough to be widely used. Metals such as iron, aluminium, zinc, and copper, can be utilized in the elemental form but most elements are so reactive that they normally exist as combinations with one or two other elements in small, simple molecules. Sodium chloride (common salt) and calcium hydroxide (slaked lime) are examples of these compounds.

Carbon is one element, however, that can form compounds with itself and other elements that can contain from two to hundreds of thousands of atoms in one molecule. In addition, the general arrangement of carbon atoms—whether in rings or chains—and the order of assembly of different groups in the molecule give rise to different compounds. Consequently, there is an almost limitless number of compounds of carbon and it is these that are categorized as organic. As with most classifications, there is an intermediate area where compounds can be considered to be either organic or inorganic, but in general the distinction is clear.

Organic compounds differ from most inorganic compounds by having relatively low melting and boiling points. Many of the simple ones are liquids or gases, indicating that the attractive forces between the small molecules are weak. Even most large organic molecules will melt or decompose at temperatures of 300–400 °C. Because carbon can be oxidized to carbon dioxide and the hydrogen, which is usually present, to water, organic compounds are seldom as stable as inorganic materials when subjected to elevated temperatures in the presence of air. Although they may be simple molecularly, inorganics generally have strong attractive forces that result in high melting points and structural strength. On the other hand, large organic molecules, because of their low melting points, can readily be formed into desired shapes and usually are less brittle than inorganic materials.

5.2. TYPES OF ORGANIC BUILDING MATERIALS

Organic materials used on or in buildings can be classified according to their use. They include liquid coatings (paints), plastics (including textiles), sealants, and roofing materials. Wood, although often placed in a separate category, is really an organic building material. Organic materials frequently contain inorganic compounds such as pigments but the basic properties of the mixture derive from the organic matrix in which the particles are dispersed. The differences between the kinds of organic materials are principally due to the type and molecular weight of the resin or binder used.

In order to possess the properties required of a building material, the molecules of an organic compound must be of large size (1). Organic solids of low molecular weight are not useful as structural materials. Large molecules are generally polymers, i.e. they contain repeating units built up by the union of a great number of a few kinds of small molecules. If these are all of the same kind the resulting large molecule is called a homopolymer. For example, styrene forms polystyrene. When there are two or three kinds of small molecules (monomers) a copolymer is produced, for example, styrene-butadiene. Copolymers usually have more desirable properties than a physical mixture of homopolymers made from the same monomers.

Because coatings are applied as liquids but must turn into solids, the original molecular size is small to intermediate and never becomes extremely large, even when the film is completely cured. With materials of lower molecular weight, the final polymer is formed predominantly after application. If the resin is already in its polymeric form before application, the coating is called a lacquer if dissolved in a solvent, or a latex paint if dispersed in water. Because of the molecular size, coatings do not have great structural strength and are, therefore, applied to substrates.

If the molecular size is increased to obtain more resistant coatings, the viscosity of dissolved resins increases so much that the amount of solvent required to permit application results in impractically thin films. Hence it is necessary to find other methods of application; one is to melt the resin, providing it is stable to heat. Heat can be applied either to the resin or to a solution of it. In the latter case, application of the material is by hot spray, while solid resin is melted after it has been applied in the form of a dispersion or dry powder.

In addition to increasing the viscosity, raising the molecular weight leads to an increase in structural strength because of greater molecular attraction and entanglement of long chains. When the material becomes strong enough to support itself at the temperature of use, a substrate is no longer required and the material is called a plastic. Strictly speaking any material that exhibits plastic flow at normal temperatures is a plastic but the term has come to apply chiefly to those organic materials which at a suitable stage in manufacture can be moulded or cast through the use of heat, pressure or both into the desired shape. Plastics can vary in properties from being hard and brittle (unplasticized polyvinyl chloride) to soft and flexible (urethane foam and synthetic rubber) (2). Again there is no distinct line between plastics and coatings since in some cases the same resin can be melt-applied, when it is called a plastic, or applied from solution, when it is a coating. A two-component urethane coating is in many ways more closely related to the polyesters used in glass-fibre-reinforced plastics than to oil paints.

Sealants contain resins that are intermediate in molecular size in comparison with those in plastics and uncured coatings. This is because sealants must possess certain unique properties to perform their function. They must have sufficient flow to be applicable but not so much as to run out of the joints and are, therefore, more viscous than liquid coatings. Sealants must retain their extensibility, however, and so cannot solidify as do coatings. Thus less cross-linking, i.e. reactions between polymer molecules, occurs in the curing of sealants, which are most closely related to rubbers.

Until recently, organic roofing materials have been based almost exclusively on asphalt or coal tar pitch reduced to application viscosity by solvents (cold-applied or 'cut-back') or heat. Generally, two or three plys of reinforcement are used to give structural strength and the asphalt or tar base provides the waterproofing. Because they are black and absorb most of the incident radiation, these materials are degraded by sunlight and thin films rapidly check (crack in a pattern of small squares). It is for this reason that asphalt-based coatings are mainly used below ground-level as foundation waterproofers or coatings for metal. On roofs bituminous materials are applied in relatively thick films and light-coloured or white gravel is embedded in the surface to protect them from light and to reduce surface temperature by reflection. In the past few years, roofing based on liquid- or film-applied synthetic resins has been introduced. These materials can be supplied in white or light colours but to date most have exhibited undesirable dirt collection or chalking.

Wood is an organic material that has molecules of extremely high molecular weight. In addition to the normally strong attractive forces between large molecules, the hydroxyl groups on cellulose lead to hydrogen bonding between molecules which link together in bundles. The chemical groups are attractive to water as well and this accounts for the ready swelling of wood by water, but the bundles are cemented together with a more water-resistant material called lignin. Because of this structure there is no solvent for the cellulose unless it is first modified chemically. As wood is produced by a living organism that must resist certain forces during its life, it has a strongly oriented structure. This orientation leads to large differences in strength and dimensional changes between the three directions—longitudinal, radial, and tangential.

5.3. WEATHERING

The process of weathering is defined as the action of atmospheric elements in altering the colour, texture, composition, or form of exposed objects, ultimately leading to disintegration or failure to perform a function. The well-known elements of the weather are radiation, moisture, thermal conditions, and gases. The composition of these elements and their effects on organic building materials is discussed in this section.

5.3.1. Radiation

Electromagnetic radiation. Unlike mechanical vibration and heat conduction, electromagnetic radiation has the property of energy

Table 5.1. *Divisions of the electromagnetic spectrum*

Name	Wavelength range
Gamma rays	0·01–1 Å
X-rays	1–100 Å (10 nm)
Ultra-violet	
Extreme	10–100 nm
Far	100–200 nm
Middle	200–300 nm
Near	300–400 nm
Visible light	400–770 nm
Infra-red	
Near	770–2500 nm (2·5 μm)
Middle	2·5–30 μm
Far	30–300 μm (0·3 mm)
Hertzian waves	
Microwave	0·3–100 mm (0·1 m)
Radio and television	0·1–1000 m

transmission across a vacuum as well as through some materials. Irradiation occurs when a material is subjected to radiation and is measured as the amount of energy incident upon a surface of given area in a given time.

Since electromagnetic radiation is in wave form, it can be described by the wavelength or the frequency of vibration. They are related by the equation $c = \nu\lambda$, where c is the velocity of light, ν the frequency, and λ the wavelength or distance between peaks. Although frequency is more fundamental, wavelength is more easily determined and has been more commonly used. The equation shows that as wavelength decreases, frequency increases. Also, as wavelength decreases, the units for measuring it are smaller. Short wavelengths are measured in Ångströms (1 Å $= 10^{-10}$ m), middle wavelengths, for example, visible light, in nanometres (1 nm $= 10^{-9}$ m), and long wavelengths in micrometres (microns) (1 μm $= 10^{-6}$ m). The longest wavelengths, radio waves, are measured in metres but in this region frequency is also used, i.e. kilohertz and megahertz.

The complete range of electromagnetic radiation is divided into regions according to properties and to wavelengths, as shown in Table 5.1. As with most classifications, the dividing lines are not sharp and one type gradually merges into the next. This is true even with

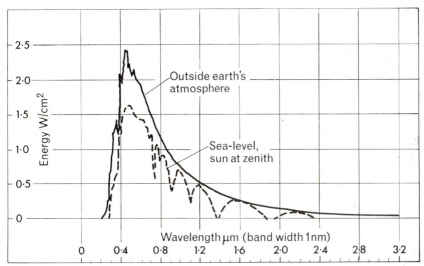

Fig. 5.1. Distribution of solar energy. Adapted from P. R. Gast, 'Solar radiation' in *Handbook of geophysics*, chap. 16–3 (New York: McMillan Co., 1960).

visibile light where the actual limits vary with the individual and his age, although the stated range is 400–770 nm. Some people can see down to 320 nm but the sensitivity is very low: at 350 nm it is only about 1 per cent of that at 400 nm (3). Hence radiation below 400 nm is called ultra-violet down to 10 nm (100 Å). Infra-red occurs beyond the visible red light, and radar and radio waves beyond the far infra-red.

Solar radiation. In the environment normal to most buildings the kind of radiant energy that has the greatest effect on organic materials is that which maintains life on earth—solar radiation. Until man starts building on the moon, he will be concerned with solar radiation as it is received at the earth's surface. Fig. 5.1 shows a comparison between the distribution of energy in the solar spectrum above the earth's atmosphere with that at sea level with the sun directly overhead at noon.

This figure shows that the solar radiation is considerably modified by the atmosphere even at the zenith. The peak in energy has been shifted from the blue to the yellow-green region and this is where the human eye, probably as a result of evolution, is most sensitive. The infra-red radiation is reduced due to absorption by water vapour and oxygen. Finally, and of greatest importance, only near ultra-violet is received at ground level while the sun emits down to 200 nm. This ultra-violet absorption is caused by ozone in the upper atmosphere (4). The reduction in energy is still greater when the sun, because of time of day or year or of latitude, is not at the zenith. At lower angles the radiation has to travel further through air, consequently more energy is absorbed. In addition, at lower sun angles the shorter wavelengths are scattered more so that only about half the ultra-violet comes in a straight line from the sun (5).

The importance of the absorption of the middle ultra-violet is related to one of the characteristics of radiation: the shorter the wavelength, the higher the energy content. This is shown by the formula $E = hc/\lambda$ where E is the energy contained in one quanta of the radiation and h is Planck's constant (Fig. 5.2). The polymers used in organic building materials are composed of long-chained molecules with carbon-to-carbon backbones. The molecules may be bound to each other by secondary forces and entanglement alone or, depending upon the type of polymer, by chemical bonds as well. Since the primary bonds are chemical and their strength depends upon the elements involved, they can be broken by sources of energy that exceed the attractive forces between the atoms, thus disrupting the molecules. In radiation, this level of energy is reached for the carbon–carbon bond at approximately 350 nm. This is well within the range of solar radiation received at sea level. Fortunately, the proportion of shorter radiation is small. At noon in the summer, ultra-violet provides 5–7 per cent of the total energy while biologically active ultra-violet (less than 320 nm) is about 1 per cent. These proportions decrease before and after noon and in winter because of the geometric effects already mentioned. For example, at 40° N. latitude, the four winter months November to February supply only about one-ninth of the biological ultra-violet provided by the months from May to August (6). In addition, clouds and smoke reduce total radiation and the intensity of ultra-violet. If it were not for these various factors, no organic polymer, including man, would have any exterior durability.

The action of radiation. When a molecule absorbs radiation it is raised to an excited state, usually at one particular atom. It may return to its

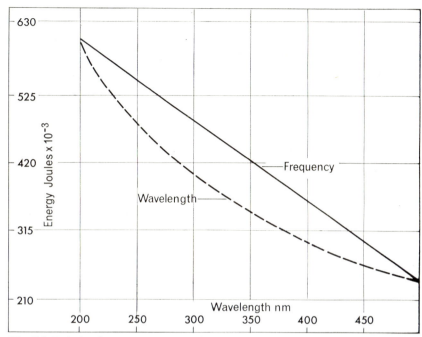

Fig. 5.2. Relation between energy and wavelength. Adapted from H. H. Jaffé and M. Orchin, *Theory and applications of ultraviolet spectroscopy* (New York: Wiley, 1962).

unexcited or ground state by dissipating the energy by reradiation of fluorescence, phosphorescence, or heat. In such a case the molecule is unaffected. This is what happens with the longer wave radiation which is turned into heat. If the radiation contains sufficient energy it may cause a chemical reaction to start at the excited atom and this frequently leads to degradation of the material.

Before either of these courses can be followed it is necessary for the radiation to be absorbed. Just because a material is irradiated does not mean that absorption necessarily occurs. Certain molecular arrangements absorb certain bands of radiation, which is why some materials are red and others blue. If none of the groups that absorb visible light is present, the material is colourless (or white if the light is reflected). This transparency or lack of absorption can also occur at wavelengths other than those in the visible region. Thus some materials may be transparent to ultra-violet but absorb visible or infra-red and vice versa. The reason that no infra-red above 2·4 μm is received from the sun at the earth's surface is that water absorbs strongly in this region although it is transparent to visible and near ultra-violet light. Similarly, ordinary window glass is opaque to that part of the ultra-violet that causes sunburn (7).

When a material is transparent to a particular wavelength it means that that radiation passes through the material without effect (8). Consequently, if a material transmits all the ultra-violet down to 300 nm, it will not be degraded by the sun. Acrylics, such as poly-methyl methacrylate (PMMA), do not absorb until well down into the middle ultra-violet and this is one reason for their excellent exterior durability. Polystyrene, however, is made from aromatic groups, which absorb at the lower end of the near ultra-violet. As a result it is affected by exterior exposure, usually turning yellow and losing some of its mechanical properties (9).

It is not sufficient for the basic polymer to be transparent. Initiators used to start the reaction must be removed after polymerization or not absorb ultra-violet. Impurities that cause absorption must also be absent, which is difficult to accomplish in large-scale production at elevated temperatures. Small quantities of absorbers can have an effect out of proportion to their amount because absorption produces chemical groups that absorb additional ultra-violet. The reaction therefore becomes faster and faster with time.

Although transparency to ultra-violet is a benefit to materials that are used in bulk, it can also be a liability if the material is a clear coating applied to a substrate which is affected by radiation. Hence, clear acrylics do not perform well on exterior wood because the top layer of the wood is degraded and the coating, being left without support, peels off. Alkyds and urethanes act in a similar manner although they absorb somewhat higher into the ultra-violet (10).

Effects on chemical constitution. Since the ultra-violet portion of radiation contains the most energy it causes the greatest damage to organic materials. The chemical degradation attributable to ultra-violet can take two paths (11). With some materials, the energy starts a process that is the reverse of the polymerization reaction that origin-ally produced the large molecules. The main chain of the polymer may be broken in isolated locations, which is called chain scission, or it may completely revert to small molecules. The latter is the so-called 'unzipping' of the polymer which fortunately occurs very slowly when radiation is the only factor. In the other process, the smaller molecules produced by chain scission or reactive sites on large molecules react with other chains. This results in more cross-linking than was originally present so that the material becomes harder and more brittle.

Chemical changes of a less destructive though still undesirable nature occur if ultra-violet alters a resin's internal structures to those that absorb blue visible light. The reflected light then appears yellow and this is generally undesirable. Another visual defect can be caused by ultra-violet even though the polymer itself may be resistant, if the material is coloured but the colorant, many of which are organic, is not. Fading, which is usually not acceptable commercially, then occurs.

As only ultra-violet possesses sufficient energy to break the primary bonds, the only chemical effect that visible and infra-red radiation have is to speed up the rate of reactions that may be occurring from other causes. The quantity of heat in solar radiation is not sufficient to raise the temperature to where chemical bonds can be broken thermally (12).

Effect on physical properties. The radiation-induced changes in the physical structure of organic materials result, of course, from the chemical reactions that have taken place. Because large molecules are required for a material to have the desired physical characteristics, if scission reduces the molecular weight too much the material soon loses these properties. If radiation causes marked depolymerization catastrophic failure occurs. For example, polymethylstyrene, which should be less brittle than polystyrene, has not been of commercial importance because it slowly reverts to monomer when irradiated by ultra-violet of 280 nm at room temperature. At 115 °C the effect of ultra-violet is seventy times greater than at 25 °C but without ultra-violet there is no degradation at all at the higher temperature (13). This illustrates the synergism common in degradation processes.

If cross-linking occurs, either from chain scission or at active sites, the material is affected because of the relation between cross-link density and physical properties. Whether cross-linking takes place is largely dependent upon the chemical structure of the polymer. Some chemical groups hinder the reaction so that there is a tendency for unzipping with consequent softening or liquefaction. Others promote cross-linking with the opposite effect: the material becomes harder.

If some flexibility is required for the building material to perform its function, the additional hardness causes cracking. Thus some coatings become less extensible and crack when they can no longer accommodate movements of the substrate. If sealants cross-link too much they either crack or lose adhesion at the interface.

In some cases the initial irradiation products are coloured and absorb subsequent ultra-violet light thus preventing deeper penetration of the radiation. Since only the outer layer becomes cross-linked, the cracking is restricted to that region. When the material has a thick cross-section, as do many structural plastics, the process results in surface cracking or crazing, depending upon the depth of the cracks. Polystyrene exhibits this type of behaviour upon exterior exposure. With coatings, which are relatively thin, the cracking may be deep enough to penetrate to the substrate. Pigmentation, however, may restrict cross-linking to a very thin layer and if the cracks are microscopic they result in chalking from the gradual erosion of this layer. The cross-linking process is more common in the degradation of organic building materials resulting from solar ultra-violet than is the depolymerization reaction.

Visible and infra-red radiation only affect the physical properties indirectly by raising the temperature. The relationship between higher temperatures and physical properties is discussed in the following section.

5.3.2. Thermal conditions

Thermal aspects of the weather relate to the presence or absence of heat—high or low temperature—and to the rapidity of change from one condition to the other: thermal shock. Temperature can affect materials both physically and chemically.

Mechanical properties and temperature. The effect of temperature on the physical properties of organic building materials may be generalized as shown in Fig. 5.3. The property illustrated here is modulus

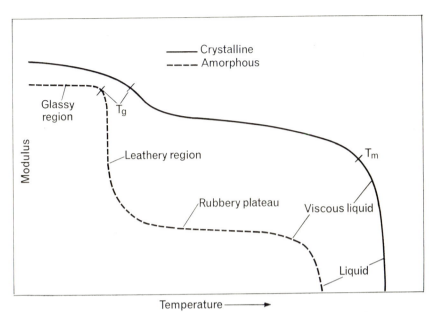

Fig. 5.3. Generalized diagram—effect of temperature on organic building materials.

(stress divided by strain) but related properties such as hardness and tensile strength exhibit a similar pattern. Elongation, creep, and flow behave in the opposite manner, i.e. they increase with increasing temperature.

The diagram indicates how the mechanical behaviour varies as the temperature increases or decreases. This type of reaction to temperature change is characteristic of molecularly complex materials. Simple compounds do not follow such curves, for example, water goes directly from the brittle state (ice) to a low viscosity liquid.

With organic building materials, the size of any one zone in relation to the change in temperature depends upon the molecular structure of the polymer. A material may remain in one region over a considerable temperature range but pass rapidly through the next one because it does not possess the structure necessary to have the properties of that region. In addition, the position of room temperature on the scale varies with the material. With some, it is in the region on the left, showing that a fairly high temperature is required before flow takes place. With others, it is towards the right, which means that low temperatures are necessary before the material becomes hard and rigid. The important point to note is that the properties do vary with a change in temperature. A common lecture demonstration of this is to cool a rubber ball in liquid air and then shatter it with a blow from a hammer.

The curves in Fig. 5.3 represent two types of polymers—amorphous and crystalline. The latter have a less complex behaviour because the strong forces between crystalline regions keep such a material solid, although it may be flexible, until it melts or decomposes. Amorphous materials, which are more common, pass through several regions as the temperature changes.

The plateau on the left of the diagram is the glassy region. In it, both kinds of polymers are hard with little extensibility (elongation) but generally have high tensile strength. Amorphous polymers in the glassy state are also brittle and shatter easily. The first inflexion in both curves is called the glass or second-order transition, usually designated by T_g. This transition zone is of greater importance with amorphous than with crystalline polymers. The latter are fairly rigid above or below their T_g so there may only be a difference in stiffness (14). The situation is quite different with amorphous materials. Several of their properties differ markedly above and below the transition point. These include thermal conductivity, specific heat, specific volume, coefficient of expansion and, of greatest significance, tensile properties. Above the T_g, amorphous polymers are no longer brittle nor are they hard as shown by the large decrease in modulus. During this abrupt change in modulus they do not possess elastic properties. Instead, with the increase in flexibility, they are said to exhibit leathery characteristics. When the transition temperature is determined in the reverse direction by cooling, it is referred to as the brittle point because of this change in properties. The method of determination can affect the result by several degrees (15). Because of the large change in many characteristics, the glass transition should not take place at the temperature of use.

As the temperature is raised higher there is another region where modulus of amorphous materials changes only slightly. This is called the rubbery plateau. For compounds of high molecular weight, the extent of the region depends largely upon the degree of cross-linking between molecules. With highly cross-linked polymers the region is small and decomposition frequently occurs due to the high temperature required to place them in this state. An example is the highly vulcanized (cross-linked) rubber—ebonite. Such materials are classified as being thermosetting, i.e. once cured they are infusible. Polymers with little or no cross-linking quickly pass through the rubbery region and become viscous liquids. They are said to be thermoplastic because they flow when heated above their softening point. Materials with the right degree of cross-linking remain in the rubbery region over a fairly wide range in temperature. If this coincides with the temperature of use, the materials are called rubbers or elastomers.

In the rubbery region, polymers exhibit long-range (up to 1200 per cent) reversible extensions with relatively small applied forces. Elastic recovery occurs because of the coiling action of long chain molecules and the occasional cross-links between them which prevent permanent deformation. The glass transition of rubbers must be well below the temperature of use to allow the molecular movements to take place (16). By contrast, the rubbery region for polymers, which are mostly crystalline, is very limited as the strong forces between crystallites maintain a high modulus that suddenly decreases at the melting point (T_m) (17).

After the rubbery plateau, there are regions of viscous and liquid flow indicated by the fall-off in modulus with increasing temperatures. The degree of cross-linking in most elastomers is sufficient to prevent much flow before thermal decomposition begins. Polymers that melt without degradation, the thermoplastics, have extremely high viscosities owing to their high molecular weight. Hence, in extrusion and moulding, pressure is used to bring about flow in a short time.

Behaviour of building materials. Coatings are applied as liquids, generally through the use of solvents instead of heat. They must, therefore, be in the liquid flow region at application temperatures. In curing, most coatings pass through the rubbery stage to the leathery region. If the glass transition of the cured film is too close to the service temperature, the material becomes brittle with only a minor or no reduction in temperature; the coating is said to lack flexibility. Most coatings are formulated so that low temperatures do not affect them particularly unless subjected to impact when very cold. Neither are the upper temperatures experienced in weathering high enough to cause softening or flow. Because coatings are applied in thin films, thermal shock is not usually important.

Sealants are in the viscous flow zone when they are applied and after curing must remain in the rubbery region at working temperatures. Hence, there must be sufficient cross-linking so that a moderate increase in temperature will not return them to the flow region. Too much cross-linking, however, would place them on the left of the plateau. With such materials a marked lowering of the temperature, which occurs in northern countries, would cause transition to the glassy state in which they are unable to elongate when the jointed units contract. Failure then occurs either by cracking of the sealant or loss of adhesion at the interface. Rate of temperature change is also important in joint movement because organic polymers, in general, can accommodate slow rates of strain much more readily than fast rates. With sealants, the adjacent building units may tend to moderate thermal shock by responding more slowly to rapid temperature changes.

Plastics intended for use in buildings must not be softened by higher ambient temperatures. Consequently, materials that must retain their form should be below their glass transition temperature if amorphous, while crystalline polymers should be well below their melting temperature (18). Because amorphous polymers must be in the glassy region they are brittle and have little impact resistance. Examples are polystyrene and polymethyl methacrylate. Where toughness is a prerequisite, polymers with a high degree of crystallinity should be used. Because both types are already in the solid state, low temperatures have little additional effect except to make crystalline polymers somewhat stiffer if they fall below their T_g. Hight emperatures reduce the ability of thermoplastics to support a load because of increased flow (creep).

Plastics of thick cross-section can experience internal stresses from thermal shock owing to the low rate of heat transmission of most organic materials. When there is a sudden temperature change the outer surface responds quickly to the new conditions while the inner portion is still at the original temperature. This can lead to surface cracking if the exterior is rapidly contracting while the interior is expanded, or to interior cracking under the reverse conditions.

Repeated thermal cycling can cause exudation of some of the less-efficient plasticizers. A whitish layer forms on the surface and what appears to be fading caused by ultra-violet light is in reality exudation resulting from repeated thermal shocks.

Temperature extremes also place demands upon asphalts and tar roofing materials. If a hard bituminous substance is used so that it will not flow in the summer sun, it may become brittle and crack badly at low temperatures. Conversely, materials that do not become too hard in winter generally flow in summer. Thus the selection of the proper grade for roofing requires considerable care. Because of the need to resist flow at high temperatures, which on a black surface can reach 200 °F (93 °C), bituminous roofing materials can only withstand low temperature shrinkage that is uniformly distributed. This emphasizes the importance of roofing design. Moderate flow at high roof temperatures is designed to repair small cracks caused by low temperatures. Rubber is sometimes added to asphalt in an attempt to overcome the problem of flow *vs* brittleness. Unless a truly homogeneous blend is obtained, the 'rubberized' asphalt will not show an improvement in ductility (19).

Chemically, the effect of temperature is to change the rate of reactions. The usual approximation is that a rise of 10 °C doubles the reaction rate. Thus oxidation, which takes place slowly at room temperature with most materials, is a much more rapid process at elevated temperatures.

5.3.3. Moisture

Water is one of the most prevalent elements of the weather. It occurs as rain or snow, water vapour (humidity), and condensed water vapour (dew or frost).

Because most organic building materials are hydrophobic and not porous, they are not so readily damaged by the freezing action of water as are many inorganic materials. Wood, being composed of a hydrophilic polymer in cellular form, is easily swollen by water. The polymer, however, neither dissolves in nor reacts with water, so that wood does not disintegrate when swollen, even if later frozen. Water is necessary for the degradation of wood by micro-organisms, even though one type is called 'dry' rot.

Some organic coatings intended for use on wood, particularly oil paints pigmented with zinc and titanium oxides, swell markedly when immersed in water (20). When water (from either the exterior or the interior of a building) collects at the back of such a paint film, the film expands in area more than the corresponding substrate and is forced off the surface, with resultant blisters. Water can also cause blistering of coatings that swell only slightly if more moisture collects at the interface than can be transmitted through the film. When the hydrostatic pressure exceeds the adhesive strength of the film, blistering occurs. Swelling properties, permeability, and adhesion to moist surfaces have, therefore, been considered important parameters in the assessment of exterior coatings for wood. Moisture can cause degradation of coatings on metal if it can permeate the film and initiate corrosion, the products of which can disrupt the coating.

Roofing materials, which are used for their waterproofing ability, and plastics are generally little affected by water. Glass-reinforced plastics can be damaged if the fibres are too close to the surface or the resin weathers away, allowing water to wick along the fibres (21). Since glass is a hydrophilic inorganic material, it is attractive to water and the reinforcing action is greatly reduced. Frozen water in the form of hail can damage brittle plastics by impact. Sealants in the bulk are also unaffected by water but their adhesion can be destroyed if water attains access to the interface. Such failures occur most often on porous surfaces that can absorb water.

5.3.4. Gases

The normal atmospheric gases that play a part in the weathering process are oxygen and carbon dioxide. The movement of the atmosphere—wind—may also be considered a weather factor. Gaseous pollutants such as ozone, sulphur dioxide, and oxides of nitrogen can cause great damage but it is open to argument as to whether they should be included in a discussion of natural weathering.

The atmospheric gas most damaging to organic materials because of its high concentration and reactivity is oxygen. Chemical linkages which are not completely 'saturated' or satisfied (chemically called double bonds) are particularly susceptible to oxidation. Indeed, this is the basis of the drying mechanism of oil paints and other coatings that cure through oxidative polymerization. Since it is impossible to have a binder that contains the exact number of double bonds to cause solidification but no more, the reaction continues past the optimum stage and becomes part of the degradation process. Hence oil paints, which depend solely upon this drying process, are more susceptible to continued oxidation than coatings which include other methods of polymerization, e.g. alkyds.

Natural and many synthetic rubbers contain unsaturation and consequently oxidize, leading to discoloration, hardening, crazing, and finally cracking. Unsaturation, however, is not essential for oxidation; polymers that contain reactive hydrogen atoms are also attacked: polystyrene (22) and polypropylene (23), for example. Since oxygen must diffuse into the material to continue the reaction, oxidation often occurs only at the surface unless the material is in a thin film.

While carbon dixode affects several inorganic building materials because of carbonation, it does not react with those made from organic materials. In fact, carbon dioxide and nitrogen (the major gaseous constituent), are frequently used in chemical syntheses as inert atmospheres. Rapid movement of the normal atmosphere—wind—can cause weathering by impinging rain, sand, or dust upon exposed surfaces. Degradation of coatings is usually more severe on the sides of buildings that bear the brunt of storms.

Ozone is normally present only in the upper atmosphere (24) so it may be considered a pollutant at ground level. Being an unstable modification of oxygen containing three instead of the normal two atoms, it is extremely reactive. Materials that oxidize will therefore react readily with ozone. An illustration is ozone cracking of rubbers. Sulphur dioxide, present in industrial atmospheres from the burning of sulphur-containing fuels, is the other common pollutant. Its action is to form sulphuric acid which may diffuse through organic coatings and attack the underlying metal.

5.3.5. Combined weather elements

Weathering is not usually the result of the action of only a single element of the weather. Generally at least two and, frequently, three factors are present at the same time, The effects of the different factors are not usually additive, i.e., a combination produces a greater effect than the sum of the individual effects. This synergism, or reinforcing action, has been demonstrated many times in studies on durability. Hence the combined actions as well as the individual processes must be appreciated, the latter being a necessary first step before the former can even be attempted. Only in this way can an adequate understanding of the weathering process eventually be achieved.

Radiation and water. Fortunately, these two elements tend to operate at different times, because the sun is not usually shining when it is raining and vice versa. However, materials can be irradiated after having been wetted by rain or when they have a high moisture content from overnight high humidity. Time of wetness, therefore, is an important parameter in relating climatic conditions to exterior degradation. The action of the combined elements can follow several paths with radiation accelerating the effect of water or the converse.

Most organic building materials have to be resistant to attack by water at normal temperatures. It is possible, however, for radiation to raise the temperature to the point where solution or hydrolysis can occur (25). Thus, plasticizers for vinyl coatings and plastics may be removed if they are appreciably soluble in water at elevated temperatures. Strength of polyester laminates can also be reduced through attack by water either on the resin itself or on the bond between the resin and the glass fibre. These actions are not as marked as in actual immersion in hot water but can contribute to the degradation process.

When coatings containing drying oils are irradiated by ultra-violet, both cross-linking and scission of the oil chains occur. The low-molecular fragments from scission could act as plasticizers but they are removed by water, consequently adding to the brittleness caused by cross-linking. The loss in flexibility then leads to cracking, which allows water to reach the substrate, resulting in loss of adhesion. This is why cracking and peeling generally occur together in coating failure. Radiation can cause cross-linking of sealants and the resultant loss of extensibility may lead to cracks between the building component and the sealant. Water can then attain access to the interface producing a further reduction in adhesion. Leaching by water of irradiated lignin is responsible for the greying of exposed wood (26).

Radiation and oxygen. A natural weathering combination that has probably an even greater effect is oxygen and radiation, referred to technically as photo-oxidation. Materials subjected to oxygen are degraded much faster in the presence of radiation than in its absence and vice versa. For example, discoloration of polystyrene occurs more rapidly when irradiation takes place in air or oxygen as shown in Table 5.2. It can be seen that the oxygen content of the product is about ten times higher when both elements operate together. In another experiment, rubber absorbed 5 per cent oxygen in the dark without major changes, but when rubber containing only 2 per cent oxygen was irradiated, the elongation at break decreased from 1100 per cent to 10 per cent (27). In a third case, the light from a 150-watt bulb accelerated the absorption of oxygen by natural rubber (28). With saturated polymers there is little damage from oxygen at room temperature if ultra-violet is absent.

Table 5.2. *Photo-oxidation of polystyrene*

Oxygen pressure (*mm*)	Ultra-violet radiation time (*h*)	Oxygen in product (%)	Colour
0	0	0·11	Nearly colourless
0	250	0·13	Light yellow
20	0	0·10	Nearly colourless
20	250	1·4	Yellow-orange

Total exposure time 250 hours at 115–20 °C in all cases; National Bureau of Standards, 'Polymer degradation mechanisms', NBS Circular 525 (1953), 206–14.

Because of the synergistic action of radiation and oxygen, anti-oxidants as well as ultra-violet absorbers are generally added to plastics designed for exterior use. Such a combination of additives has been shown to be more effective than ultra-violet absorbers alone in preventing yellowing of polystyrene by radiation, provided the anti-oxidant does not absorb ultra-violet. If it does absorb radiation, photochemical degradation is accelerated (27), emphasizing the care needed in selecting additives.

Irradiation in space might be expected to cause greater damage to organic materials owing to the presence of shorter wavelength ultra-violet. With some polymers the effects are actually less deleterious because of the absence of oxygen. In a vacuum they cross-link rather than depolymerize (29), so that while the material may become brittle, it does not evaporate. With such resins it has been possible to produce coatings that maintain the temperature of satellites within design limits.

Radiation and heat. The radiation referred to in this combination is again ultra-violet since infra-red radiation is converted to heat. Degradation reactions that proceed only at temperatures higher than those reached on natural exposure may occur at much lower temperatures under the influence of ultra-violet. Thus, polystyrene when heated in a vacuum remains colourless as it depolymerizes at 320 °C but pre-irradiated polystyrene rapidly turns yellow, then brown, and finally black at considerably lower temperatures (30). Polyvinyl chloride also is degraded faster thermally after or during irradiation. Plastics that have been held at high temperatures too long during moulding are more susceptible to ultra-violet degradation upon exposure.

Oxygen and heat. The effect of temperature on oxidation has been briefly referred to. Polystyrene again can be used to illustrate the synergism; in an inert atmosphere it begins to degrade at 280 °C but, in air, slow oxidation takes place at 110 °C and it burns at 245 °C. Similarly, styrene-acylonitrile copolymer can be heated for an hour at 200 °C in a vacuum without changing colour, but in air it becomes yellow and partially depolymerizes (31). Oxidation of polyethylene and polypropylene is negligible at low temperatures but a moderate increase in temperature markedly increases the absorption of oxygen. Although this example illustrates the effect of a combination of factors, it is not a practical problem with polyethylene because it melts before becoming hot enough for rapid oxidation to take place. With other plastics also, the temperatures experienced in weathering are, fortunately, not sufficiently high to cause drastic oxidation. Nevertheless, there is an effect on the service life of materials that tend to oxidize.

Other combinations. The total number of combinations possible with four factors is 15, i.e. $(2 \times 2 \times 2 \times 2) - 1$, if absence of all four is not considered to be a weather condition. When the possibility of different levels of each element is introduced it can readily be seen why studying the effects of the weather is such a difficult task. It is, therefore, impracticable to discuss each imaginable set of factors and only a few of those remaining will be mentioned.

Water and oxygen are both essential in the rusting of iron which can disrupt organic coatings or other building materials that tend to

Table 5.3. *Degradation* of polyvinyl chloride*

Exposure condition		Atmosphere	
Radiation	*Heat*	*Nitrogen*	*Oxygen*
None	1 h, 150 °C	0·0035	0·0043
Sun lamp	1 h, 150 °C	0·009	0·021

*Measured by amount of HCl produced.
National Bureau of Standards, 'Polymer degradation mechanisms', NBS Circular 525 (1953), 91.

restrict the expansion of rust. Moisture and heat are each more damaging in the presence of the other. When held at 70 °C, nylon becomes brittle after two years in a dry atmosphere, but in only eight weeks under wet conditions (32). Plastics softened by heat are more readily eroded by wind-driven particles such as sand. On the other hand, wind can reduce the temperature of irradiated surfaces. Ozone cracking occurs much more extensively when the material is under mechanical stress (33).

The degradation reactions taking place when three elements of the weather are present at the same time are even more complex. As expected, three factors acting together are more damaging than any combination of just two. This is illustrated in Table 5.3 for heat, radiation and oxygen.

5.4. REDUCING THE EFFECTS OF THE WEATHER

Because the effects of combined elements are so severe, damage to organic building materials might be mitigated if one of the factors could be eliminated or at least minimized. The examples given in Tables 5.2 and 5.3 show that radiation tends to be the most important of the three factors of heat, oxygen, and radiation. Although heat can be controlled, one cannot normally remove oxygen or water from the environment on the exterior of most buildings. Consequently, many attempts have been made to diminish the radiation, there being three basic approaches that can be taken to achieve this objective.

The first method of minimizing the effect of radiation is to use polymers that do not absorb it. As previously discussed, it is difficult

Fig. 5.4. Silicone polymer structure

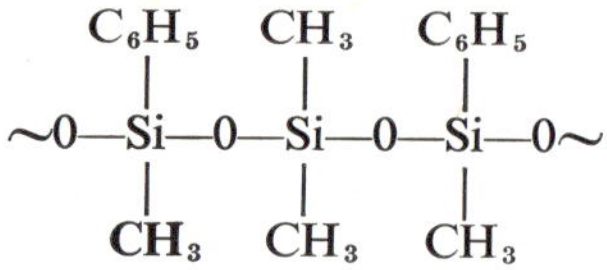

$$
\begin{array}{ccc}
\mathrm{C_6H_5} & \mathrm{CH_3} & \mathrm{C_6H_5} \\
| & | & | \\
\sim\!\mathrm{O}\!-\!\mathrm{Si}\!-\!\mathrm{O}\!-\!\mathrm{Si}\!-\!\mathrm{O}\!-\!\mathrm{Si}\!-\!\mathrm{O}\!\sim \\
| & | & | \\
\mathrm{CH_3} & \mathrm{CH_3} & \mathrm{CH_3}
\end{array}
$$

Fig. 5.5. Fluorocarbon structures.

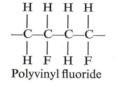

Polyvinyl fluoride

Polyvinylidene fluoride

Polytetrafluoroethylene

to achieve complete transparency except on the laboratory scale. Thus, polyethylene, which should be transparent to ultra-violet, readily degrades on exterior exposure (34).

The second method would be to make polymers from combinations of elements whose bond strengths exceed the energy available in solar radiation. Unfortunately, most such combinations form simple compounds instead of polymerizing. Of the few polymers that can be made, many are readily decomposed by water or oxygen. Hence the possibilities of success with this approach are limited. The best-known polymers of this type are probably the silicones in which there is a silicon–oxygen backbone with organic side-groups, as shown in Fig. 5.4. The silicon–oxygen bond is only broken by radiation of wavelengths below 270 nm and this is not received at the earth's surface. The organic groups are necessary for the material to have the properties required of a polymer; without them the material is quartz or silica—SiO_2. Fluorocarbons are another example of this kind of polymer. Although fluorine is not part of the molecular backbone, the high strength of the fluorine–carbon bonds in the side-groups (Fig. 5.5) contributes markedly to their excellent durability (35).

The final and most common procedure in reducing the effect of radiation is to prevent the polymer from absorbing it. If the material does not have to be transparent, this can readily be accomplished through the incorporation of pigments that reflect radiation or absorb it preferentially. Reflection usually occurs at the pigment surfaces within the resin so that the radiation has to pass through the top layers twice. Some degradation can, therefore, occur at the surface, and this is why materials frequently lose gloss on exposure. For complete absorption to take place the pigment must be black. Incorporation of black pigment is very effective as shown by the increase in durability of polyethylene from one year to twenty years with the addition of 1 per cent carbon black (36). The colour, however, is not always acceptable. For other colours titanium, zinc, or iron oxides can be used, but higher concentrations are required.

If the polymer must be clear, preventing it from absorbing radiation is much more difficult. It is then necessary to use compounds which are, in effect, dyes that absorb ultra-violet but not visible light. In order to continue providing protection, however, the compound itself cannot be destroyed but must dissipate the energy harmlessly as heat. Since it is already a problem to obtain dyes that are stable to visible light, it is much more difficult to find ones that are resistant to the higher energy ultra-violet radiation. The few compounds that do perform satisfactorily are referred to as ultra-violet absorbers or stabilizers. These materials are most efficient when used in materials that have a thick cross-section because the amount required is a function of concentration and thickness. Thus, a plastic 0·020 in (0·5 mm) thick might be stabilized with 0·5 per cent absorber but would require 1 per cent at 0·010 in (0·25 mm) and 2 per cent at 0·005 in (0·125 mm). This relationship is not strictly linear; effectiveness is reduced at higher concentrations so that more is required than calculated from the formula. Clear coatings which are applied at 0·001 in (0·025 mm) per coat require an absolute minimum of 5 per cent to be effective on exterior wood in two or three coats. Unfortunately, this quantity almost doubles the cost of the raw materials and it becomes economically prohibitive. In addition, at high concentrations compatibility of the ultra-violet absorber and the polymer being protected may become a problem.

Another complication is that the compounds are rather specific in their action; even absorbers that are closely related chemically may show large differences in effectiveness with different resins (37). As a result, comprehensive tests are needed to determine the type and amount of absorber to be used with any given polymer. Finally, it should be appreciated that the absorbers do not last indefinitely but are slowly degraded. Hence the absorption they are intended to prevent will ultimately occur.

5.5. SUMMARY

The effects of the different elements of the weather on organic building materials have been described in this article. Because of the complexity of the over-all weathering process, most of the discussion is devoted to how the materials react to the individual elements. This approach is considered necessary in developing an understanding of the weathering of organic building materials.

The four chief weather factors are radiation, temperature, water, and oxygen. It was shown that radiation tends to have the greatest effect and that ultra-violet light between 300 and 350 nm is the most damaging part. Ultra-violet acts by changing the chemical structure of the polymers used in organic building materials, consequently affecting their physical properties and their ability to perform a given function. Fortunately, the intensity of the destructive ultra-violet is a small part of the total solar radiation and is reduced as the angle of the sun decreases and by clouds and smoke. Still, enough is received on the surface of buildings to cause degradation of materials, particularly when the radiation acts in conjunction with oxygen, water, heat, or a combination of these elements.

The severity of damage is increased with combinations of factors. Whether preventive action is required to reduce the degradation to an acceptable rate depends chiefly upon the ease and cost of replacing or repairing the material. Thus for a single-family dwelling, original and renewed application of the normal types of coatings is usually considered to be more economical than initial use of extremely durable but expensive coatings that require extensive surface preparation and great care in application. In the future this situation may change as the cost of labour increases in relation to the cost of materials. With high-rise buildings, it is necessary that the original materials retain their appearance and ability to perform the desired function for as long as possible. To this end, some of the steps taken to lessen the effect of radiation on organic materials include production of polymers unaffected by ultra-violet and incorporation of compounds that reflect or harmlessly absorb ultra-violet. Admittedly, the preventive measures are normally in the hands of the material manufacturers. Nevertheless, building designers, constructors, and users should be aware of their importance and their limitations in improving the performance of organic building materials.

REFERENCES

1. MILLER, M. L. (1966). *The structure of polymers*, p.3 (New York: Reinhold).
2. SAVETNICK, H. A. (1966). 'The plastic materials', in Skeist, I. (ed.), *Plastics in building*, chap. 2 (New York: Reinhold).
3. KOLLER, L. R. (1965). *Ultraviolet radiation*, p. 7 (New York: Wiley).
4. —— Ibid., p. 118.
5. LUCKIESH, M. (1946). *Germicidal, erythemal and infrared energy*, p. 52 (New York: Van Nostrand).
6. —— Ibid., p. 55.
7. KOLLER, L. R. Loc. cit., p. 227.
8. KING, A. (1968). 'Ultraviolet light: its effect on plastics', *Plastics and Polymers*, **36** (123), 195.
9. BUCKNALL, C. B. (1966). 'Degradation and weathering of polystyrene and styrenated polyester resins', in Pinner, S. H. (ed.), *Weathering and degradation of plastics*, chap. 6, p. 87 (Manchester: Columbine Press).
10. ASHTON, H. E. (1967). 'Clear finishes for exterior wood, field exposure tests', *J. Paint Technology*, **39** (507), 212.
11. JELLINEK, H. H. G. (1967). 'Fundamental degradation processes relevant to outdoor exposure of polymers', in Kamal, M. R. (ed.), *Weatherability of plastic materials*, p. 42 (New York: Interscience).
12. ROSATO, D. V. (1968). 'Radiation', in Rosato and Schwartz, R. T. (eds.), *Environmental effects on polymeric materials*, chap. 7, part II, p. 672 (New York: Interscience).
13. STOKES, S., and FOX, R. B. (1962). 'Photolysis of poly-α-methylstyrene', *J. Polymer Science*, **56** (164), 507.
14. BRYDSON, J. A. (1966). *Plastics materials*, p. 42 (London: Iliffe).
15. —— Ibid., p. 37.
16. LENZ, R. W. (1967). *Organic chemistry of synthetic high polymers*, p. 28 (New York: Interscience).
17. —— Ibid., p. 33.
18. BRYDSON, J. A. Loc. cit., p. 57.
19. BARTH, E. J. (1962). *Asphalt science and technology*, p. 625 (New York: Gordon and Breach).

20. Browne, F. L. (1957). 'Swelling of paint films in water: XI Mixed-pigment paints in linseed oil', *Forest Products Journal*, **vii** (7), 248.
21. Crowder, J. R. (1965). 'The weathering behaviour of glass-fibre reinforced polyester sheeting', *BRS Miscellaneous Papers*, no. 2 (July).
22. Grassie, N. (1966). 'Fundamental chemistry of polymer degradation', in Pinner, S. H. (ed.), *Weathering and degradation of plastics*, chap. 1, p. 5 (Manchester: Columbine Press).
23. De Paolo, P. A., and Smith, H. P. (1968). 'New phenolic phosphite stabilizers for polypropylene', *Stabilization of polymers and stabilizer processes*, chap. 15, p. 204 (Washington: American Chemical Society).
24. Koller, L. R. Loc. cit., p. 120.
25. Hauck, J. E. (1966). 'How water affects plastics', *Materials in Design Engineering*, **64** (6), 93.
26. Frey-Wyssling. (1950). 'Discoloration of unprotected wood by the weather', *Schweiz. Zeit. Forstw.* no. 101, p. 278.
27. Field, J. E., Woodford, D. E., and Gehman, S. D. (1955). 'Infrared study of oxidation of elastomers', *J. Polymer Science*, **xv** (79), 51.
28. Stafford, R. L. (1948). 'The oxidation of rubber in light', *Proc. Second Rubber Technical Conference*, p. 94.
29. Koller, L. R. Loc. cit., p. 259.
30. Grassie, N. Loc. cit., p. 8.
31. Bucknall, C. B. Loc. cit., p. 90.
32. Harding, G. W., and MacNulty, B. J. (1961). 'The embrittlement of polyamides: its mechanism and preventive treatment', *Society of Chemical Industry Monograph*, no. 13, p. 395 (London).
33. Zuev, Y. S. (1953). 'Protection of rubber from ozone cracking', *Doklady Akad. Nauk SSSR*, no. 93, p. 483.
34. Brydson, J. A. Loc. cit., p. 78.
35. —— Ibid., p. 205.
36. Heaps, J. M., and Austin, A. (1966). 'Degradation of polyolefins', in Pinner, S. H. (ed.), *Weathering and degradation of plastics*, chap. 7, p. 113 (Manchester: Columbine Press).
37. Philadelphia Society for Paint Technology. (1967). 'Ultraviolet light absorbers in clear coatings for wood', *J. Paint Technology*, **39** (515), 736.

6

Non-destructive testing of concrete

R. H. Elvery, BSc (Eng), CEng, FICE, MI Struct E
Senior Lecturer, Department of Civil and Municipal Engineering, University College, London

J. A. Forrester, BSc
Chief Chemist, Cement and Concrete Association, Wexham Springs, Slough, Bucks

6.1. INTRODUCTION

During the last three decades investigators in several countries have examined the possibilities of using various testing techniques to assess the condition of concrete without affecting the subsequent behaviour of the samples tested.

Such non-destructive tests have many attractions and could be useful for a range of practical applications from the routine of quality control to special investigations both in the laboratory and on the construction site.

Efforts to establish valid methods of testing concrete non-destructively have been helped to some extent by the experience gained from similar activity in the testing of other materials. Some of the methods applicable to these materials cannot be used on concrete because its structure is essentially different from that of most of these other materials. However, a body of knowledge exists relating to the technology of the non-destructive testing of wide range of materials and it is becoming increasingly clearer which parts of this technology can be reliably applied to concrete testing.

Non-destructive testing could, by definition, include simple operations such as visual examination, the measurement of density by weighing, and the measurement of deformation under limited stress. It could also include special techniques requiring the use of equipment such as X-ray or ultrasonic apparatus. It is now generally accepted that the term 'non-destructive testing' refers to such specialized methods and the discussion in this article is confined to this latter meaning.

6.2. THE NEED FOR TESTING CONCRETE NON-DESTRUCTIVELY

Before considering the various methods which are available for testing concrete non-destructively, it is as well to discuss the reasons why these types of test are of practical importance.

Concrete consists essentially of a mixture of cement and water added to either naturally occurring or to artificial mineral aggregates and these form a solid conglomerate when the cement paste hardens as a result of the hydration of the chemical compounds that make up the cement.

The ability of the hardened concrete to resist deformation or disruption when mechanical forces are applied to it or to endure the attacks of aggressive environmental conditions, depends on a number of different factors..These factors can be controlled in varying degrees and it is possible to select both the proportions and types of the constituent materials to produce concrete having the required properties within acceptable limits under particular conditions of manufacture.

These limits can sometimes be relatively wide and it is not always economically feasible to improve their precision. It is, however, important that the extent of the variations be known so that they may be taken into account when the concrete structure or product is designed and specified. It is therefore necessary to establish both suitable definitions for the properties required and reliable means of measuring these properties to check compliance with the specified requirements.

The traditionally accepted testing procedures for concrete include standard tests of the cement, aggregates, and water and also the analysis of the freshly mixed constituents to check composition. They also include mechanical tests on samples cast from representative batches of the concrete used for the construction of the structure or product.

In Britain the samples most commonly used are cubes of 4 in or 6 in side and these are tested to destruction by applying a compressive load to two opposite faces of each cube in a direction at right-angles to that in which the concrete is cast. The corresponding test specimen used in the USA is a cylinder 6 in dia. $\times$ 12 in high and the loading is applied in the same direction as that of casting. To a lesser extent prisms (either 4 in $\times$ 4 in $\times$ 20 in or 6 in $\times$ 6 in $\times$ 28 in) are used as samples and these are subject to flexural tests to determine the modulus of rupture. Other tests include the measurement of density (gravimetrically) and the absorption of water by the concrete after a suitable period of drying.

All of these and most of the other standard tests on concrete are concerned with checking various stages in its manufacture and none of them is designed to assess the state of the actual concrete in the structure or product after all the stages of manufacture have been completed. The mechanical tests on cubes, cylinders, or prisms indicate the potential resistance of the concrete to special arbitrary loading conditions after being thoroughly compacted and stored in favourable conditions of temperature and humidity.

These tests may be used to provide a measure of quality control for the manufacture of concrete but are not designed to check fully the quality of the finished concrete in the structure or product. In particular they leave unchecked two important stages of manufacture, namely the compaction of the concrete and its subsequent maturity.

All tests on hardened concrete whether they are destructive or not are necessarily in the nature of post-mortem examinations since some time must elapse to allow the concrete to harden. It is then usually too late to take any effective action to alter the condition of the concrete that has been tested. Cube tests, for example, are usually made when the concrete is 7 or 28 days old, and by that time, a substantial volume of concrete will have been manufactured before the cube test results are available.

Some types of non-destructive tests enable useful information to be obtained at early ages and may also be used to monitor the progress of the maturity of the concrete as time progresses. This may be used to advantage to enable formwork to be struck or prestressing operations to be carried out as early as possible. This facility for testing and retesting the same piece of concrete at any convenient stage in its life can be used also to determine changes in its structure due to the influence of environmental conditions which may cause it to deteriorate.

Most structural concrete is either reinforced with steel bars or is prestressed with high tensile steel tendons embedded in it. Assessment of the quality of structural members must take into account the interdependence of both steel and concrete whether they are reinforced or prestressed.

The mechanical strength of reinforced concrete members is often more dependent on the quality and position of the steel than on the quality of the concrete (within certain limits). On the other hand the concrete is required to protect the steel from corrosion and it can only do this satisfactorily if its depth of cover and its resistance to moisture penetration are adequate. Thus a reliable assessment of the quality of a reinforced concrete member requires a check on the position of the steel as well as on the quality of the concrete.

In prestressed concrete members precision of positioning the prestressing tendons is very important and so also is the protection of these tendons by the surrounding concrete or by injected grout (in the case of post-tensioned members).

Non-destructive testing techniques are available which can be used to check the density and elastic stiffness of the concrete, the position of reinforcing bars and prestressing tendons and the effectiveness of grout injection into prestressing ducts. They can also provide useful information to enable the seriousness of defects due to poor workmanship or to environmental action to be assessed more reliably than by visual examination.

A particular advantage of non-destructive testing is that regions of a structure selected for their critical importance may be examined in detail. As a sampling procedure, this is a considerable improvement on the usual method of random selection of samples for destructive tests. In view of the significantly wide quality variation found from point to point in a structure, these samples are hardly likely to be very representative of the concrete in the critical regions.

Apart from field applications, these techniques fulfil many of the special needs met with in laboratory investigations especially those concerned with the examination of structural changes within concrete which may occur with time due to various causes. Their use can exploit the facility of retesting individual specimens many times and this enables more information to be obtained for a given number of specimens as compared with the adoption of destructive methods.

To sum up, traditional destructive test methods, while providing adequate and necessary checks on various stages of concrete manufacture, fall short of providing data to enable the service behaviour of concrete structural members to be reliably assessed. There is, moreover, a need to assess the seriousness of damage or deterioration which structures may have suffered under service conditions. There is a need to provide a better feedback of test information to allow effective action to be taken to influence subsequent production. The specific needs for information include the determination of homogeneity, density, elastic stiffness, and position of reinforcement. Most of these needs can be met by the proper application of available non-destructive test methods.

6.3. APPLICATIONS OF NON-DESTRUCTIVE TESTING

Fig. 6.1 shows the relative components that go to make up the state of hardened concrete that can be examined non-destructively. We shall deal with each one in turn.

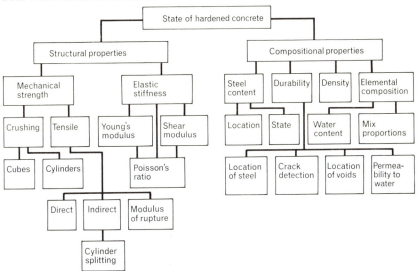

Fig. 6.1. Range of information on the state of hardened concrete obtainable from non-destructive tests

6.3.1. Elastic stiffness

Young's modulus. This quantity is required to determine the elastic deformation of structures for design calculations and its measurement is also a useful check on the homogeneity and development of stiffness of the concrete as it matures or deteriorates. It can be determined as the dynamic modulus of elasticity on sample pieces and on small precast units by measurement of the longitudinal or flexural resonance frequency. The geometry must be limited to certain shapes and the density must be known as will be explained later.

The dynamic modulus can also be determined from ultrasonic pulse velocity measurements when Poisson's ratio and the density of the concrete are known. It can also be determined empirically directly from pulse velocity measurements.

Shear modulus. This quantity is not frequently required but the dynamic shear modulus may be determined on small specimens by the measurement of torsional resonance frequency. As for longitudinal resonance, the geometry and density of the concrete must be known. It can also be determined from the ultrasonic shear wave velocity if the density of the concrete is known.

Poisson's ratio. This may be required for structural analysis purposes and it can be derived from a determination of longitudinal resonance frequency together with either torsional resonance frequency or longitudinal ultrasonic pulse velocity.

6.3.2. Mechanical strength

Both compressive and tensile strengths are normally measured by destructive tests on specimens loaded to failure. Any attempts to determine strength non-destructively must necessarily depend on there being some empirical correlation between strength and some type of non-destructive test.

Correlations have been found between ultrasonic pulse velocity or dynamic modulus determined from resonance frequency measurement, and both compressive and tensile strengths. Many factors influence these relationships and they must be studied empirically.

Similarly empirical relationships are also found to exist between mechanical strength and surface hardness and again a number of different influences must be taken into account.

Combined measurements of ultrasonic pulse velocity and surface hardness have been found (1) to sharpen up the empirical correlation with strength as have also combinations of pulse velocity and pulse attenuation. Correlation with strength is particularly important to engineers but the definition of strength deserves to be considered with some care as will be discussed later.

6.3.3. Compositional properties

Density. This can be determined to indicate the degree of compaction of the concrete and also to assess its homogeneity. It can be measured by radiometric techniques using either X-rays or gamma-rays.

Elemental composition. This can be related to the specified mix proportions of the concrete and can be used to show variability in its composition. The techniques for measuring water content of the hardened concrete include neutron moderation and microwave absorption. Techniques such as activation analysis are not very well developed but will show calcium and silicon contents of concrete. When relating these to the cement content of concrete, care must be taken to allow for the relative proportion of these elements in cement and aggregate.

6.3.4. Durability

Density as affected by porosity and lack of compaction can be an important indication of durability of concrete and this has been discussed above.

Crack detection. Surface cracks can be seen visually with or without the aid of a microscope but the use of fluorescent ink enables very narrow cracks and microcracks to be shown up vividly by a rapid and simple process.

The depth of relatively pronounced cracks which can be seen at the surface can be determined with reasonable accuracy from ultrasonic pulse measurements.

Permeability. This can be assessed non-destructively with the initial surface absorption test (6).

Location of steel. Steel reinforcement which is too near the surface of the concrete may corrode and cause cracking and subsequent loss of durability to the concrete. The depth of concrete cover can be measured by magnetic methods or by gamma radiography.

Voids in concrete. Honeycombing in concrete due to poor compaction can be seen by gamma radiography and can also be detected by ultrasonic pulse tests.

Voids in the grout inside ducts for prestressed concrete cables can be filled with air or water. Where steel cables pass through grout and air, corrosion is likely to occur at the interface. When the ducts have water-filled voids, these can expand on freezing with consequent damage to the concrete. Such voids, whether filled with air or water, can be seen by gamma radiography.

Deterioration due to sulphate attack or frost action. This can be studied on laboratory specimens by measurement of the reasonance frequency. The extent of damage developed on structures in service can be determined by using the ultrasonic pulse technique.

6.3.5. Steel content

Bar diameter and state of steel. Checking bar diameters for compliance with design requirements may be done using gamma radiography. The same method, if used with care, can give some indication as to the extent of severe corrosion of the steel bars.

Location. The position of reinforcement or prestressing cables is an important factor in determining the strength of structural members. This may be checked by gamma radiography, radiometry, or by magnetic methods.

6.4. THE AVAILABILITY OF NON-DESTRUCTIVE TESTING TECHNIQUES

The technical and scientific literature contains an extensive collection of reports on the development and use of a variety of techniques for testing concrete non-destructively. Much of the work described in these reports has been exploratory or has been concerned with special laboratory investigations and this has been reviewed at various times (2–4). From this work there has emerged a number of test methods which are readily available for practical use either in the laboratory or in the field.

However, in the UK the techniques are used to a limited extent only and are frequently considered to be more useful for 'trouble-shooting' rather than for routine inspection. This is to be regretted since many of the methods could be made more effective if their use were planned in advance of construction rather than as an after-thought or as a last resort to track down the extent of defects that become visually apparent.

One probable reason for this situation is that the equipment available for some of these tests is usually in a form more suitable for laboratory use than for site use. This view is supported by the fact that, for instance, the rebound hardness test is relatively more popular than other non-destructive tests because of its simplicity, low cost, and ease of operation. In spite of these advantages, the validity of the test results obtained with it is more limited than that of some of the other methods.

There is also some reluctance to give non-destructive methods general recognition by official bodies or by structural designers. Nevertheless, there is currently a renewal of interest in their potentialities. Recently the British Standards Institution have issued BS 4408 (5) relating to some of these tests and drafts of further specifications are at present under discussion.

The range of techniques which have been studied to explore the bases of possible non-destructive techniques is shown diagrammatically in Fig. 6.2. From this range, a number of useful methods have been developed and these are to be found in the four main areas into which the diagram is divided. Before discussing which areas of the diagram are useful for providing various kinds of information, the physical bases of the four main areas should first be considered.

6.5. VIBRATION METHODS

6.5.1. Application of vibration theory

When the theory of sound is used to examine the behaviour of stress waves in a solid, it is evident that the structure of the solid has a considerable influence of this behaviour. It would therefore appear that the measurement of certain behaviour parameters when concrete is subjected to mechanical vibrations might provide significant information about its structure.

Most of the theoretical studies of stress waves in a solid have considered the solid medium to be isotropic, homogeneous, and perfectly elastic. From these studies it is clear that the velocity of mechanical vibrations is influenced by the mode of vibration and the density and elastic stiffness of the material. The velocities of different modes of vibration are dependent on different elastic constants.

For example, the velocity of a longitudinal wave in an elastic solid of semi-infinite dimensions depends on the density, Young's modulus,

P.C.S.—13

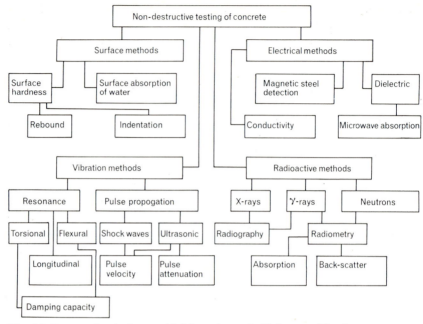

Fig. 6.2. Range of techniques which have been studied or used for the non-destructive testing of concrete

and Poisson's ratio of the solid. If, however, the same type of wave is propagated along a thin rod of the same material, its velocity depends only on Young's modulus and density. The theoretical relationships concerned with these cases are given in equations [1] and [2].

Longitudinal wave velocity (m/s) in a semi-infinite medium

$$= V_1 = \sqrt{\left\{ \frac{E}{\rho} \frac{(1-\nu)}{(1+\nu)(1-2\nu)} \right\}} \quad [1]$$

and longitudinal wave velocity (m/s) in a long thin rod

$$= V_2 = \sqrt{\frac{E}{\rho}} \quad [2]$$

where E = Young's modulus (N/m²),
ν = Poisson's ratio,
ρ = density (kg/m³).

Evidently, the measurement of wave velocity in a rod enables Young's modulus to be determined provided the density of the material is known. Also if the wave velocity is measured in both rod and a large block of the same material the value of Poisson's ratio may be determined from equations [1] and [2].

There are two possible ways of measuring the velocity of vibrations in a solid, namely, either be determination of the resonant frequency of the sample or by timing the flight of short pulses of vibrations passing through the sample.

6.5.2. Resonance methods

The measurement of resonant frequency is readily applicable to a rod and this can be done by causing the rod to vibrate in longitudinal mode at its natural, resonant frequency. Under these conditions, a standing wave is set up and it, can be shown that the wavelength of the longitudinal vibrations is equal to twice the length of the rod. Thus if the frequency of the vibration at resonance is measured the wave velocity may be determined from the relationship:

$$V_2 = \lambda n = 2nl \qquad [3]$$

where λ = wavelength of vibration (m),

l = length of the rod (m),

n = resonant frequency (Hz).

Then from equations [2] and [3]

$$2nl = \sqrt{\frac{E}{\rho}}$$

or $E = 4n^2l^2\rho$.

This method of measuring Young's modulus, described in BS 1881 (6) as a 'test for the dynamic modulus of elasticity by an electrodynamic method', is generally more sensitive than the ultrasonic pulse test described in the next section. It is possible to measure the resonance frequency of specimens vibrated in torsional or flexural modes and the corresponding elastic constants may be devised. These techniques are restricted to testing discrete specimens since the relationship between resonant frequency and the properties of the material depends very much on the size and shape of the specimen tested.

Resonance tests are therefore quite unsuitable for *in-situ* testing and cannot be used to examine local areas of concrete in a structure as the ultrasonic method can. However, for laboratory specimens and some types of precast units, resonance tests are particularly useful because of their sensitivity since this enables very small changes in concrete properties to be measured reliably.

6.5.3. Pulse methods

The measurement of pulse velocity requires electronic timing apparatus to measure the time taken for the leading edge of a short train of waves to travel between two points in a material which are at a known distance apart. Such a pulse may be generated by a sharp blow from a hammer or by using an electro-acoustic generator.

When a mechanical impulse is applied to a point on the surface of a solid material, three distinct kinds of energy radiation occur. These are known respectively as longitudinal, shear, and surface (or Rayleigh) waves.

The longitudinal wave has particle displacement of the material in the same direction as that in which it travels. It travels about twice as fast as the other two types of wave.

The shear wave is the next fastest and this has a particle displacement at right angles to the direction of travel. The Rayleigh wave travels along the surface and has an elliptical particle displacement.

The use of electro-acoustic transducers to generate pulses enables better control of the type, directionality and frequency of the pulse to be obtained than when hammer blows are used. Generally, electro-acoustic transducers are designed to produce longitudinal pulses since these give more useful results than the other types of pulse. Even so, some of the energy is propagated as shear and Rayleigh waves but the faster longitudinal pulse has a leading edge which becomes clearly separated from these and its exact time of arrival can be definitely identified without interference from them.

In general, the velocity of a pulse travelling through a solid medium is not the same as that of a continuous wave of the same type travelling in the solid. However, if the solid is in the form of a large block, both waves and pulses travel with the same velocity and only when the dimensions of the block in a direction at right angles to that of propagation (i.e. the lateral dimensions) become relatively small do their velocities differ. The wave velocity is reduced as the lateral dimensions decrease but the pulse velocity is reduced to a much smaller extent.

Thus pulse velocity is almost independent of the shape of the piece of material through which it passes and is dependent on the elastic properties of that material. This makes the measurement of pulse velocity a very useful test method for *in-situ* structural concrete with the range of sizes and shapes encountered in practice.

The behaviour of pulse transmission through a solid is affected by the frequency of the waves comprising the pulse although the pulse velocity is not influenced except when the lateral dimensions of the solid are small in relation to the wavelength of the pulse. When the frequency of the pulse is very high the energy is directed through the material within a narrow beam but high frequency pulses become attenuated more than low frequency pulses as they pass through the material.

In metals, this attenuation is small and it is preferable to use very high frequencies (i.e. more than about 1 MHz) to take advantage of high directionality in order to detect and locate small flaws in the material. This is usually done by measuring the time taken for a pulse to travel to the flaw and return as an echo.

In concrete it is not possible to use an echo method in this way since every boundary between an aggregate particle and the cementing matrix generates an echo which is indistinguishable from that sent back by a flaw. This behaviour reduces the directionality of high frequency pulses since the multiple echoes generated within concrete result in a diffusion of the vibration energy which is scattered in all directions.

The use of pulses for testing concrete is therefore more successful when these are of relatively low frequency since directionality is of less importance than it is in metal testing and attenuation is considerably reduced. Nevertheless the frequency range of pulses suitable for concrete testing is still beyond the range of human audibility and is said to be *ultrasonic*.

The range is from about 20 kHz to about 200 kHz, a compromise being required between the high frequency end, which provides a pulse with a very sharp leading edge but is rapidly attenuated, and the low frequency end which has a less well-defined leading edge but can travel long distances in concrete without losing most of its energy.

In practice the measurement of pulse velocity is made by recording the time taken for a pulse to travel from a transmitting transducer,

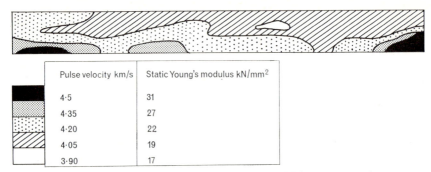

	Pulse velocity km/s	Static Young's modulus kN/mm^2
	4·5	31
	4·35	27
	4·20	22
	4·05	19
	3·90	17

Fig. 6.3. Ultrasonic pulse velocity variation in a typical laboratory-made reinforced concrete beam. Structural members made in the field usually show a wider range of variation than this

through a known length of concrete to a receiving transducer, the transducers being placed at convenient positions on the concrete surface. The electronic equipment is designed to record the time at which the leading edge of the first pulse reaches the receiving transducer. Satisfactory pulse transmission requires the provision of good coupling conditions between the concrete surface and each of the transducers. This is usually ensured by applying grease or some similar coupling medium to the face of each transducer.

The most effective positions of the transducers for good transmission of energy are at points directly opposite to each other on different sides of a piece of concrete since the energy propagated is still significantly directional. There is, however, sufficient energy propagated parallel to the surface of the concrete to allow detectable transmission to occur between two transducers placed on the same concrete surface.

Generally the ultrasonic test measures the pulse velocity in the concrete which lies directly between the two transducers so that it is particularly useful for exploring the state of the concrete between two opposite faces. This can provide a method of scanning a region of a concrete member to determine the variations which occur from point to point right through that member. A typical example of such an exploration is shown in Fig. 6.3 which indicates the variations in a reinforced concrete beam.

When a void or large discontinuity occurs in the direct line between the two transducers, the pulse travelling along this line is interrupted

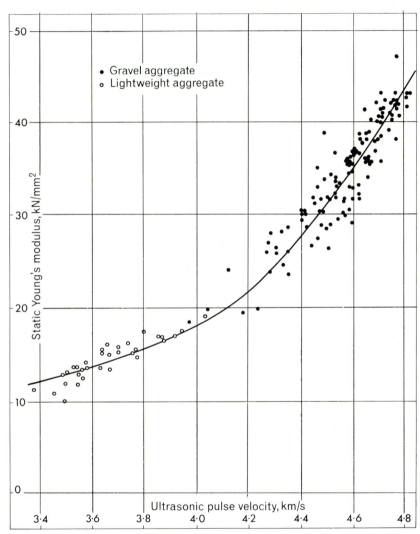

Fig. 6.4. Experimental correlation between ultrasonic pulse velocity and static Young's modulus. Results of experiments by Evans (7) and Baron-Hay (8)

and does not reach the receiver. The receiving transducer detects the leading edge of the fastest pulse which has taken a route around the void. The equipment then records a longer transmission time than it would have done if no void had been present and this enables the void to be located.

When the transducers are both located on the same surface, the pulse velocity measurements obtained generally relate only to the surface concrete and give no information about the heart of the concrete member. This technique is, however, useful for estimating the depth of cracks visible at the surface.

6.5.4. Empirical relationships

So far this discussion has considered the use of vibrational methods as a means for the determination of wave or pulse velocities and the use of the results to calculate properties such as elastic stiffness. In practice there are generally two parameters which are of particular interest to the structural engineer, namely Young's modulus and mechanical strength.

The values of Young's modulus determined from vibrational tests (i.e. the dynamic modulus) usually differ from those obtained from static load-deflexion tests mainly because concrete does not behave as a perfectly elastic material. This difficulty can be overcome by using empirical relationships between the results of static and dynamic tests. Experiments have shown that there is a good correlation between static and dynamic moduli for a wide range of concrete types.

A particularly useful empirical relationship is that between static modulus and ultrasonic pulse velocity since this holds good for a range of conditions and types of concrete. Fig. 6.4 shows typical experimental results for concrete made with two very different types of aggregate, an artificial lightweight aggregate and a natural river gravel. This relationship enables the value of static Young's modulus to be determined at any point in a concrete structure where it is possible to measure the pulse velocity.

The mechanical strength of concrete can also be related empirically to Young's modulus but this requires careful consideration of the choice of an appropriate definition of strength. This will be discussed later in this article.

6.6. RADIOACTIVE TECHNIQUES IN NON-DESTRUCTIVE TESTING OF CONCRETE

Techniques based upon the use of X-rays or gamma-rays and neutrons reveal properties of materials at the atomic level and measurements are unaffected by the way the constituent atoms are bound together. It is possible using X-rays and gamma-rays to measure density of concrete and to illustrate by radiography areas of concrete which are less dense by virtue of poor compaction or voids and areas of greater density such as are due to the presence of steel. Using fast neutrons it is possible to measure the total water both bound and free present in the concrete. The techniques are non-destructive and non-residual.

It should be possible to relate density measurement made by radiation absorption and water content by neutron retardation to non-destructive methods of measuring structure sensitive properties such as pulse velocity and resonant frequency methods. However, relationships between results from radioactive techniques and those determining structural properties of the concrete such as cube crushing strength and modulus of elasticity can only be empirical because structural properties depend upon crystal properties and inter-particulate forces.

Some work has been done using nuclear radiation to measure the constituents of hardened concrete but this again relates to the elemental composition and where an element can occur in the aggregate and the cement, for example, calcium and silicon, the technique must be augmented with information on the individual constituents.

6.6.1. Principles

To understand the use of radioactive techniques we must first look at some principles of nuclear physics. An atom consists principally of a central nucleus containing protons and neutrons around which orbit a number of electrons. Protons are particles of unit mass with a unit positive charge. Neutrons are particles of similar mass with no electrical charge. In an element the number of protons present in the nucleus is the atomic number and the sum of the protons and neutrons constitute the atomic weight. With the exception of hydrogen the nuclei of the elements in concrete contain almost as many neutrons as protons, i.e., the ratio of atomic weight to atomic number is nominally two. Hydrogen contains one proton and no neutrons and the ratio of atomic weight to atomic number is one. In an atom there are as many electrons as protons. Electrons have little mass ($\frac{1}{1850}$ of unit mass) and carry a unit negative charge. Isotopes are atomic species of the same elements with the same atomic number which defines their chemical behaviour, but different atomic weights which define their nuclear behaviour. The unit of energy used in nuclear physics is the electron volt (eV) which is the energy acquired by an electron in falling through 1 volt (1 eV $= 1\cdot6 \times 10^{-19}$ J).

Nuclear radiation is energy transmitted either by the motion of particles such as electrons (called beta particles), alpha particles, protons, neutrons, and other less known particles, or by short wavelength, high frequency electromagnetic radiation such as X-rays and gamma-rays. X-rays which originate in the electron cloud surrounding the nucleus start at a frequency of 3×10^{16} Hz (cf. visible light around 5×10^{14} Hz) and gamma-rays, which originate in the nucleus of the atom, range around 3×10^{18} Hz.

At these energies electromagnetic radiation is a series of discontinuous packages of energy known as quanta or photons. The quantity of X-rays and gamma-rays is measured in röntgens (r). (One röntgen will produce one electrostatic unit of electricity in 1 cm^3 of dry air at $0\,°C$ and 1 atmosphere pressure.)

Radioactive elements emit nuclear radiation when the nucleus disintegrates to a more stable form. Heavy, naturally radioactive elements emit mainly alpha particles and artificially produced radioisotopes (which are elements that have been bombarded with elementary particles) emit mainly beta particles and gamma-rays. Disintegration is random (that is any one nucleus is just as likely to disintegrate as another) but the number of disintegrations during a time interval is a measure of radioactivity. The unit of activity is 1 curie (Ci) ($= 3\cdot7 \times 10^{10}$ disintegrations per second). Disintegration results in a loss of radioactivity called the 'decay' of the radioactive material and the half-life of a radioactive source is the time taken for its radioactivity to drop to half the original level.

The various types of nuclear radiation differ in their ability to penetrate matter. Beta particles are absorbed by a few millimetres of

Table 6.1

Source	Gamma energy (MeV)	Half-life	Optimum working thickness of concrete for radiography (mm)	Dose-rate at 1 m (rad/h Ci)*
Cobalt-60 ^{60}Co	1·13 and 1·33	5·26 a	150–450	1·30
Caesium-137 ^{137}Cs	0·66	30 a	100–300	0·33
Iridium-192 ^{192}Ir	0·296 to 0·613	74 d	30–200	0·48

*rad is the unit of absorbed dose. One rad equals 0·01 J/kg.

aluminium while a much greater thickness of lead is needed to arrest gamma-rays. Neutrons are unaffected by electrostatic forces as they have no charge and their penetration, being arrested only by collision with a nucleus, is high.

6.6.2. Density measuring techniques

When concrete is placed in the path of X-rays or radiation emitted by gamma sources the radiation is scattered or absorbed in a manner dependent upon the density of the material. This technique is described as direct transmission and has been used on roads (9), concrete cores (10), walls, beams (11), and structures (12–15), and for finding the bulk density of concreting materials (12). It has been the subject of a recent detailed study by Preiss (16).

When concrete is brought into contact with a gamma emitting source and detector arrangement where the detector is screened from direct radiation from the source, some of the radiation scattered from the concrete is reflected to the detector. With certain provisions the backscattered radiation can be related to the density of the concrete. This technique is described as 'backscatter' and has been used on soils (17–19) and concrete or cement-stabilized soils (17, 20–3). The geometries of these two techniques are shown in Figs. 6.5 and 6.6.

A second type of backscatter geometry has been described (24–5), where the scattering of gamma-rays through 180° is measured. Here there is no shielding of the source from the detector and the quantities of scattered and direct radiation detected are separated electronically.

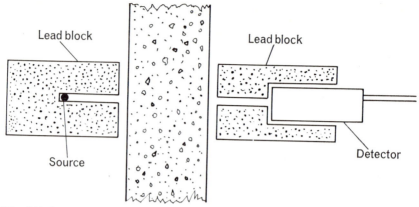

Fig. 6.5. Geometry of direct transmission gamma radiometry

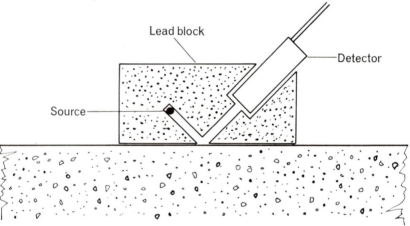

Fig. 6.6. Geometry of backscatter gamma radiometry

The sources used for transmission and backscatter techniques are usually cobalt-60 or caesium-137. The detectors are either Geiger–Müller tubes or scintillation counters. A Geiger–Müller tube is a closed tube filled with gas which becomes ionized and produces a pulse when a gamma photon enters the tube. These tubes produce electrical pulses of constant height regardless of gamma ray energy and they take up to 400 μs to register the photon.

Scintillation counters are more recent and are based upon the ability of a phosphor crystal to convert into light emission some fraction of the energy lost during the passage of a gamma photon through the crystal. These scintillations are counted by a photo-multiplier and result in electrical pulses which can be amplified and counted. These counters have a short resolving time and the electrical pulse height is proportional to the photon energy. Pulses from Geiger–Müller tubes and amplified pulses from scintillation counters are counted by scalers for a given time or by ratemeter. The pulses from scintillation counters can be analysed by a pulse height analyser to distinguish photon energy.

6.6.3. Theory of the direct transmission technique

In transmission geometry the ideal situation is either to collimate the source and detector to eliminate scattered radiation or to discriminate electronically in the detector against scattered radiation whose energy is lower than the mono-energetic radiation from the source. Such radiation would have been reduced in energy by scattering. In this ideal situation the relationship between the incident radiation intensity (I_0) and the detected radiation intensity (I) can be expressed as a Poisson distribution function,

$$I = I_0 \exp(-\mu t) \qquad [4]$$

where t is the thickness and μ is the linear absorption coefficient.

The linear absorption coefficient which varies with photon energy and the nature of the concrete and may be derived from

$$\mu = \mu_m \rho$$

where μ_m is the mass absorption coefficient and ρ is the density.

The mass absorption coefficient can be the sum of four constituent absorption coefficients depending upon the absorption mechanisms. Different mechanisms predominate at different energy levels of the radiation and different atomic weights of the absorber. Of these four only two are of interest when energies of radiation less than $2 \cdot 5$ MeV are considered; these are Compton scattering and the photo-electric effect.

If a concrete contains no element with an atomic number greater than 26 and the radiation energy is between $0 \cdot 2$ and $1 \cdot 0$ MeV then we have only to seriously consider Compton scattering as the attenuating mechanism. In this mechanism the photon is scattered by collision with an orbital electron. If Z_i and A_i are atomic number and atomic weight respectively for the elements present in the concrete as mass fraction G_i, then the Compton mass absorption coefficient μ_{mc} is given by

$$\mu_{mc} = N X \xi$$

where X, the Compton scatter factor, equals

$$\sum_{i=1}^{i=n} \frac{Z_i}{A_i} G_i \quad .$$

The Compton cross-section per electron ξ is dependent only upon the photon energy and is given by Klein Nishina formula (26). N is Avogadro's number.

The Compton scatter factor is almost constant and equal to 0·5 for all elements except hydrogen for which it is 1. At each scatter the photon loses energy in being deflected through an angle θ. If E_0 (MeV) is the incident photon energy then the energy of the scattered photon E (MeV) is given by

$$E = \frac{E_0}{1 + \dfrac{E_0}{0\cdot51}(1 - \cos \theta)} \quad . \qquad [5]$$

If and when the energy finally falls below 0·2 MeV then there is a greater probability that the photon will be annihilated by the photo-electric effect. In this process the photon transfers all of its energy to an orbital electron which is then ejected as a photo-electron. The mass absorption coefficient for this process μ_{mp} is given by

$$\mu_{mp} = N Y k$$

where k is the electron cross-section for this reaction and Y is the photo-electric effect factor which equals

$$\sum_{i=1}^{i=n} \frac{Z_i^5}{A_i} G_i \quad .$$

As can be seen this process is very dependent upon the atomic number and therefore on the chemical composition of the concrete.

We therefore have a complete equation for the attenuation of radiation through concrete within the limitation imposed, i.e. gamma-rays with energies less than 2·5 MeV and the elimination of scattered radiation by collimation or by energy discrimination at the detector:

$$I = I_0 \exp\left[-\rho t N(\xi X + k Y)\right] \quad .$$

The Compton scattering factor is almost equal to 0·5 and varies no more than 3·2 per cent for extreme changes in the composition of normal concrete. The photo-electric effect factor can change by 22·6 per cent with change of composition of concrete but it constitutes only about $2·5 \times 10^{-5}$ of the mass absorption coefficient and can usually be ignored. If we consider only unscattered radiation passing through concrete we have a relationship between density and intensity:

$$I = I_0 \exp(-\rho t N \xi X)$$

or

$$\rho = \frac{\ln\left(\dfrac{I_0}{I}\right)}{N \xi X t} \quad . \tag{6}$$

6.6.4. Deviations from the ideal direct transmission state

Deviations from the ideal relationship result from detection of scattered radiation and the effects of chemical composition. In a non-ideal state where some scattering is detected an allowance must be made in the attenuation relationship, equation [4], for a build-up factor:

$$I = B I_0 \exp(-\mu t) \quad .$$

The build-up factor B can be calculated as the sum of two simple exponentials,

$$B = A \exp(-a_1 \mu t) + (1 - A) \exp(-a_2 \mu t) \quad .$$

The parameter a_1, a_2, and A have been derived empirically and for concrete and photon energy 1 MeV, $A = 8·0$, $a_1 = -0·11$ and $a_2 = 0·044$. This effect can be eliminated by collimation of the source and detector and by selecting electronically only the pulses from the detector due to direct radiation.

From the theory of operation, it can be seen that the chemical composition of the concrete can produce errors varying up to 3 per cent in the Compton scattering factor which produces corresponding errors of similar magnitude in the derivation of density. Preiss (16) has shown that this error can be almost eliminated by finding empirically an apparent Compton scattering cross-section per electron (β)

to replace the theoretical value of ξ in equation [6]. The range over which β was found to be constant included all practical chemical compositions, densities from 1600 to 2500 kg/m³ and thicknesses from 100 to 150 mm. An accuracy of 95·5 per cent confidence of 0·5 per cent in the density can be achieved by this practical technique. Accuracy is limited by the physical density measurement employed for the calibration specimens which can be affected by local heterogeneity.

6.6.5. Theory of backscatter techniques

With these techniques the theory is more complex, the principles of scattering and absorption of radiation described above still apply but the detector sees only radiation scattered from the concrete and originating within the concrete by uncollimated scattering processes.

We saw in equation [5], which was for one collision, that the energy loss and the angle of deflexion are related. When multiple collisions are considered the pattern becomes more complex. Because of the interest in the use of concrete as shielding material the complexity has been studied in some detail (26–7). A study of angular distribution and energy spectra resulting from radiation penetrating concrete has been made using the Monte Carlo method and a computer to consider the theoretical histories of 5000 photons (28). This analysis compares well with practice and shows energy and angular distributions for radiations from cobalt-60 and caesium-137. It opens the way for precise calculation of the backscatter radiation from a point source near a concrete surface to a nearby detector. A further study by Chilton (29) shows the effect of source and detection placement.

For a particular backscatter geometry the response of the detector increases with density to a maximum and then decreases with increase of density. The position of the peak depends upon source and detector geometry (23–5), source energy (29, 30), and degree of collimation (11, 28). In some gauge designs the peak occurs at a density of 1000 kg/m³ and with others the peak can be at densities greater than 3000 kg/m³ (23–5).

Very little use has been made of these theoretical studies in calibrating backscatter devices. Most calibration curves are derived empirically. The calibration curve is usually affected by chemical composition of the concrete (30, 31) particularly with gauges using the

Geiger–Müller tube or where there is no energy discrimination. This sensitivity to chemical composition reflects the dependence of the backscatter response on the photo-electric effect. This can be reduced by energy discrimination in the detector (23–5) or by the 'air gap' technique developed by Kühn (30) which can utilize the relatively cheaper Geiger–Müller tubes. In this technique a measurement is first made with the gauge on the surface of the soil or concrete. A measurement is then made with the gauge raised a predetermined height above the surface. For calibration these measurements are made on standard density blocks and the ratio of the 'gap' count to the 'contact' count or the logarithm of this ratio is plotted against the density of the blocks. The success of the technique depends on elimination in the ratio of the constituent of the readings influenced by the photo-electric effect. A further study to optimize the gap distance to reduce photo-electric effect influence but still maintain sensitivity to change of density has been made by Gardner *et al.* (32).

Without allowing for the photo-electric effect influence, the backscatter technique relies upon calibration curves prepared using material of the same composition. Using standard curves of calibration, coefficients of variation around 3 per cent have been found (19). Eliminating or allowing for photo-electric effect in the detection technique reduces this coefficient of variation to 1 per cent.

In the backscattering technique the concrete being examined must either be greater in size than the volume looked at by the gauge or allowance must be made for the material immediately adjacent to the back or side of the concrete. This volume varies in shape and size with the source of energy (30) the geometry of the gauge (23, 25) and the density of the concrete (31).

Examination shows that the effective depth of measurement is dependent on the density of the concrete. For concrete of density 2300 kg/m^3 values of around 100 mm for angular backscatter (21) and 60 mm for 180° backscatter (25) have been found. It has been shown (31) that 50 per cent of the measured radiation comes from the top 12–25 mm of concrete. This limits the assessment of the concrete density with depth but does not detract from the use of this technique to survey the change in quality over an area, as is done by an American device called a 'Road logger' which surveys at a speed of 1·3 m/s (18).

6.6.6. Determination of moisture by neutron moderation

The use of this technique in borehole exploration for petroleum was first described in 1941 (18). It has been applied as a surface gauge to soils (18), concrete and soil cement mixtures (11, 13, 18), to control water in concrete batching plants, and to measure the water content of concrete mixes (12, 33). The method does not differentiate between free and bound water and some effect of density has been noted (34).

The principle of operation is discussed in detail in a number of publications and will only be briefly dealt with here. Some radio-isotopes disintegrate with the emission of alpha radiation and when these are associated with a suitable light element target such as beryllium fast neutrons are generated. A list of these generators is given in Table 6.2.

Table 6.2. *Neutron generators*

Nuclide	Half-life	Approximate emission with Be target n/s Ci	Approximate gamma dose-rate at 1 m per 10^6 n/s Be target
Actinium—227	21·8 a	$1·5 \times 10^7$	8
Americium—241	458 a	$2·5 \times 10^6$	1
Curium—242	163 d	$2·5 \times 10^6$	1
Polonium—210	138·4 d	$2·5 \times 10^6$	$<0·1$
Radium—226	1620 a	$1·3 \times 10^7$	60
Thorium—228	1·913 a	$2·0 \times 10^7$	30

When fast neutrons are introduced into a material they collide with the nuclei of the atoms. Some are absorbed into the nuclei producing radioactive elements whilst others are scattered elastically and non-elastically. For the lighter elements elastic collision predominates and the energy of the neutrons is reduced. Successive collision reduces the neutron energy to that of the surrounding atoms and it becomes a thermal neutron.

The most efficient element for slowing down fast neutrons is hydrogen because its mass is about the same as that of the neutron. Though they have some effect (34) the elements other than hydrogen contained in soil and concrete are considerably less efficient moderators of fast neutrons than hydrogen. Therefore, the hydrogen contained in the water in soils and concrete is almost solely responsible for neutron moderation by the soil or concrete and a measure of the thermal neutrons produced is a measure of the water content.

Thus, in the technique, if a fast neutron generator and a detector for thermal neutrons is placed adjacent to soil or concrete the response of the detector will be in proportion to the water content. Other hydrogen containing material such as organic compounds will affect the result as will thermal neutron absorbers such as boron. The most common detector used is a boron trifluoride tube which is similar to a Geiger–Müller tube but filled with boron triflouride gas. Recently, thermal neutron sensitive phosphors have been produced and these allow the use of scintillation counters.

Where the source is also a gamma emitter it can be used in a combined density backscatter and moisture meter. The accuracy is usually of the order of a 2 per cent coefficient of variation and surface gauges appear to measure the water content to a depth of 125–50 mm (31).

6.6.7. Radiography by X-rays and gamma-rays

When concrete is irradiated from a point source by X-rays or gamma-rays the effects of the transmitted radiation can be seen by its effect on a suitably sensitive film. Areas of the concrete less dense due to voids will show as dark areas due to reaction of the film with greater radiation intensity and conversely steel will be seen by corresponding light areas on the film.

The application of X-radiography to concrete was first described by Mullins (35) but the limitation of energy from portable X-ray sets makes this technique only suitable for thin sections on site. Gamma-ray emitting isotopes with their greater energy were first used to locate steel in a concrete slab by Whiffen (36) and the potentialities of gamma radiography of concrete were soon explored for locating voids, steel (37, 40) and grouting defects in post-tensioned, prestressed concrete (38). The present state of the art has recently been described (39) and a British Standard on the technique has been published (5). A typical radiograph is shown in Fig. 6.7.

The isotopes commonly used for radiography are listed with their properties in Table 6.1. As the source cannot be switched off, suitable containers must be used which allow the source to be shielded when it is not in use. Provision is either made for a closeable aperture to allow

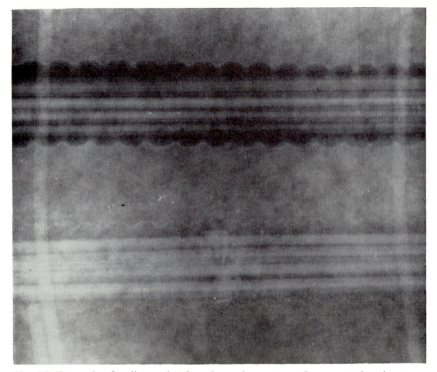

Fig. 6.7. Example of radiograph taken through prestressed concrete showing two prestressing ducts, one grouted and one ungrouted

the release of a well-defined divergent beam or for the source to be transferred from the container by remote control through a flexible pipe to a prepared position. Modern source holders made of spent uranium allow the use of a cobalt-60 source up to 15 Ci activity. These will radiograph up to 460 mm of concrete within 1 h.

A good radiograph must have clear definition. The contrast between light and dark areas increases with exposure but a limit to exposure is imposed by ability to view the radiograph when light from a viewer is directed through it. Penumbral effects which result from the finite size of the source can cause lack of sharpness of the image in a radiograph. The geometric unsharpness factor U is defined by,

$$U = \frac{dT}{f-T}$$

where d is the projected source diameter,

T the maximum object to film distance, and

f is the source to film distance.

U should be kept small and this is done by using small diameter sources, putting the film as near the area of interest as possible and making the source to film distance as large as is practicable.

The exposure time in hours is given by $S\,F\,f^2/A$ where A is the source activity (Ci), S is a concrete thickness factor which relates empirically to concrete thickness, and F is a film sensitivity factor on an arbitrary scale which ranges from 1 to 20 as shown in Table 6.3.

Table 6.3. *Film factors*

Manufacturer	Film type			
Kodak	Kodirex 1·0	Industrex D 2·3	Crystallex 5·0	
Ilford	G 1·3	B 2·6	CX 5·8	F 14·3

There is much experience available in the use of this technique and commercial radiographers are available who can carry out an investigation.

6.6.8. Other radioactive techniques

The technique of activation analysis has been used to measure the elemental composition of cement (42), soil–cement mixtures, and concrete (43). Here the material is bombarded with thermal neutrons and the gamma-ray spectrum produced is analysed for radioactivity produced. Bombardment of calcium-48, present in natural calcium, by thermal neutrons produces calcium-49 which has a gamma energy of 3·1 MeV that can be isolated by a gamma-ray spectrometer. Silicon content can be analysed similarly from fast neutron bombardment. Here aluminium-28 is produced and its discrete gamma-ray energy can be measured with a spectrometer.

These techniques measure calcium or silicon in the concrete as a whole and for a determination of cement content either the cement or the aggregate must be analysed separately. Calcareous or siliceous aggregate interferes with the cement content determination from calcium or silicon respectively.

6.6.9. Safety aspects

Radioactive materials and radiation have long been recognized as potential sources of injury to body tissue and research has been directed to find out how much radiation can be tolerated with safety. In the UK the use of ionizing radiation comes within the scope of the Factories Act and is controlled by the Ionizing Radiations (Sealed Sources) Regulation 1969 (41). Under this regulation maximum doses of radiation received by persons connected or unconnected with the work with the sources are laid down.

Gamma-ray sources must be stored in containers that shield the radiation and reduce its level to the recognized safe value. Neutron sources are stored in blocks of paraffin wax and if they are gamma emitters then additional shielding must be provided for the radiation.

6.6.10. Advantages and disadvantages of radioisotope techniques

The techniques described have all been developed to be within the limits of accuracy required for practical application. They are free from errors of human judgement and are non-destructive. They consider only small increments of the whole, therefore they are useful for defining variability and they are quick to perform. A conventional density and moisture test takes about 20 min for two technicians, whereas with the nuclear equipment one technician can do the job in 5–10 min. The operations are relatively simple and within the scope of a trained technician. There is, however, a radiation hazard and with undeveloped techniques accuracy is limited. The nucleonic apparatus is expensive and sometimes not rugged enough for rough handling.

Future development will, however, obviously correct the disadvantages and improve the advantages and the techniques have an optimistic future.

6.7. SURFACE METHODS

Surface hardness testing has been established as a number of traditional methods of inspecting localized points in metals and metal products to measure resistance to indentation and wear. These test measurements have been correlated with other engineering properties of the metal and the tests have been used for routine quality control purposes.

Hardness testing of metals has generally been done by means of indentation techniques such as the classical Brinell test or by a rebound method like the Shore Schleroscope. Similar methods may be used to test concrete surfaces and, of these, the rebound method is the most popular. It has the advantage of simplicity and is suitable for testing both precast and *in-situ* concrete.

Unlike metals, concrete is permeable to water and the degree to which water can penetrate it has an important influence on its ability to resist the action of aggressive climatic conditions or of a corrosive environment. A third surface test method for concrete has been introduced in recent years which has no counterpart in metal testing and which is designed to assess the durability of the concrete by measuring the absorption of water by its surface. This is known as the initial surface absorption test and is described in BS 1881: 1969: Part 5 (6) as well as in a number of other BS specifications for various types of precast products.

None of these tests can provide information on the state of the concrete within the heart of a structural member and the assessments they give are less comprehensive than those obtained by ultrasonic or radioactive methods. This is an important limitation particularly in view of the fact that the concrete at the surface of a structural member is usually unrepresentative of that in the main body of the member. Furthermore, the condition of the concrete at the surface is usually significantly affected by the type of formwork if it is a moulded surface or the finishing process if it is an unmoulded surface.

This limitation on the validity of the surface methods does not apply to the same extent to the surface absorption test since the state of the surface is particularly important so far as durability is concerned.

The indentation method is to apply an impact force to the concrete surface through a steel ball and to measure the diameter of the indentation formed on the surface.

The apparatus most commonly used for the rebound test is the Schmidt rebound hammer. In this a plunger is held in contact with the concrete surface and a spring-loaded mass strikes the end of the plunger and rebounds. The amount of the rebound is measured on a scale graduated in terms of arbitrary rebound numbers. The apparatus is designed so that rebound measurements can be easily made in rapid succession.

It can be argued that the rebound test results must be considerably influenced by whether the point on the surface where the instrument is applied is on a piece of aggregate or on a region of cement matrix. This is true for any individual test measurement but this difficulty can be overcome by taking the average reading for a number of tests made in the vicinity of the chosen point so that a more representative result is obtained.

Rebound tests are suitable only for close-textured surface and should not be used for no-fines concrete or for areas of poor compaction in normal concrete. In carrying out the test, the axis of the instrument is positioned in a direction normal to the surface under test. For *in-situ* testing it may thus be necessary to use the instrument in an inclined direction and calibration curves must be referred to to allow for this.

The results of rebound tests are frequently correlated with cube test results and it is found that no single correlation curve exists for this since there are a number of influential factors besides cube strength which affect rebound measurements. There is some douubt about the usefulness of such an exercise as will be discussed more fully later in this article. If such a correlation is required it is generally necessary to prepare a correlation curve for the particular concrete used. More detailed reports have been prepared by Kolek (44–5) concerning the rebound test.

The initial surface absorption test has developed from experience in the precast concrete field which indicated that a reasonable correlation exists between the amount of water absorbed by a dry piece of concrete and its durability. The test for this (i.e. the water absorption test) is made on prepared sample pieces taken from concrete units. These samples are dried in an oven and then immersed in water. The amount of water absorbed is determined by measuring the changes in weight of the samples at various intervals of time. This test is destructive and usually cannot be made on a concrete member when it is in service. It is also relatively insensitive as a means of assessing durability.

The initial surface absorption test, which has been studied by Levitt (46), may be done either in the laboratory or on the site by measuring the volume of water absorbed by a given area of the dry concrete surface, after times of exposure to a given head of water varying from 10 min to 2 h. The most satisfactory results are usually obtained on laboratory specimens which have been oven-dried. Tests carried out on the site are made on concrete surfaces which are considered to be sufficiently dry if no water has fallen on them during the previous 48 h. Such tests are less accurate than those done in the laboratory.

6.8. ELECTRICAL METHODS

6.8.1. Magnetic methods

The need to provide sufficient cover of concrete over a steel bar in reinforced concrete to maintain a passivating environment to the steel makes the measurement of this cover an important technique. The presence of the steel can be determined by the effect it has upon the field of an electromagnet (48–9).

In principle the cover meter works by measuring the reluctance of a magnetic circuit. The search unit consists of a U-shaped core of high magnetic permeability upon which are mounted two coils. Through one of these coils is passed an alternating current and the current induced in the other coil is measured. The induced current depends upon the mutual inductance of the coils and this is affected by the proximity of the steel reinforcement and the extent to which it closes the air gap. The standing current in the second coil in the absence of any reinforcement is balanced by a suitable backing-off current. The induced and backing-off currents are rectified and a moving-coil weter measures the unidirectional differential current which flows mhen the induced current increases due to the closing of the gap. The size of the bar has little effect but the concrete must have low magnetic permeability for it not to affect the reading. Depths of cover up to about 100 mm are measurable with this technique.

6.8.2. Conductivity and dielectric measurement

The simple measurement of moisture content by measurement of conductivity is limited by the concentrations of salts in solutions which occur in the water in concrete. Whereas the technique can be used on wood, with concrete this concentration effect is a normal limitation. Tobio (50) has reduced this limitation by measuring the conductivity of plaster blocks embedded in the concrete. Measurements of conductivity and capacitance are best made on samples introduced into a measuring head and they are more suited to the measurement of water in sand and aggregate. The dielectric constant of a material increases with its water content and it can be measured from a ratio of the capacity of a fixed condenser with and without the material between the plates.

6.8.3. Microwave absorption

Water is a polar liquid displayed dipolar relaxation phenomena in the energy band of centimetric radio waves otherwise known as microwaves. The state in which the water is held affects to some extent, the relaxation energy but the high absorption at this energy still occurs. As the absorption depends upon the quantity of water present, a method of assessing the moisture phase of porous material is available.

This technique has been developed and used to measure the moisture content of walls and slabs (51). The apparatus consists of two parts, a portable radio transmitter and a receiver between which is situated the material where water content is to be measured. The transmitter works at a frequency of 3000 MHz (a wavelength of 100 mm) modulated at a frequency of 3 kHz. The divergent beam having passed through the material is picked up by a receiving waveguide connected to a tuned crystal detector and amplifier. Any metal in the path markedly affects the readings. A commercial model operating at 2450 MHz is currently available. The nature of the test material and its thickness affects the readings and the accuracy on a uniform material is claimed to be ± 4 per cent of the mean value. With concrete, internal scattering and diffraction can reduce this to between ± 12 per cent and ± 30 per cent of the mean value.

6.9. IMPLICATIONS OF NON-DESTRUCTIVE TEST RESULTS

Advances in non-destructive testing techniques tend to provide greater refinements in inspection methods and these bring to light more detailed information about the product being tested. Imperfections are revealed which were previously too small to be detected and had hitherto been ignored.

This situation could cause embarrassment unless a rational assessment is made of the implications of the information revealed. If this is not attempted, the degree of sensitivity of a testing technique could become the deciding factor for acceptance or rejection in routine testing.

An ultrasonic pulse velocity scan of a structural member like that shown in Fig. 6.3 reveals a state of the concrete which is hardly familiar to many experienced concrete engineers. In this case the concrete in the regions with the darkest shading is estimated to have values of cube strength and elastic modulus which are about twice as great as those estimated for the concrete in the regions with no shading.

Such a condition must be considered as normal for concrete construction and its revelation does not make the structure any weaker than it was before we knew of the existence of the variations within the concrete. Thus the use of a technique which enables a more stringent scrutiny to be made should not necessarily lead to a more ready condemnation of the concrete construction inspected.

It is therefore particularly important to establish the practical significance of any information which non-destructive testing may provide. To do this it is useful to consider the approach of the design engineer who first attempts to establish the service performance required and then devises a structure with an estimated performance which matches these requirements. For each of these operations there exists an area of relative uncertainty in the quantitative estimates involved and this is taken care of by a judicious choice of margins of safety in each case.

The uncertainty in the estimation of strength of a structure is partly due to the variations of material quality and workmanship which are to be expected. Non-destructive tests can often establish more precisely the extent of these variations and could lead to a justifiable reduction in safety margins which would otherwise need to be rather conservative.

The present methods of design for structural concrete takes concrete strength into account by expressing it in terms of cube strength. Consequently when non-destructive testing has been used the results have generally been expressed as estimated cube strengths. Cube strength has become firmly established as a yardstick of concrete quality and it is used as a universal definition of strength almost without question. For this reason the validity of any new testing techniques have been judged in relation to their success in having a good correlation with cube strength.

Both ultrasonic pulse velocity and rebound hardness tests have been examined in this way and it has been found in each case that a family of curves is required to relate them to cube strength due to the many factors influencing this relationship. This has unfortunately resulted in an unjustifiable loss of confidence in these tests and this has delayed their general acceptance as valid methods of test.

There is, however, a growing awareness that the cube test has serious limitations and its universal correlation with the service performance of structures is now open to some doubt. There is a reasonable case for using the results of non-destructive tests as valid indications of quality in themselves without attempting to express them in terms of cube tests. In general, it is more efficient to determine the relationship between the strength of structural members and the results of non-destructive tests if these are to be used. This would provide a means of specifying concrete quality in a way that would allow a direct check to be made on the members rather than on test pieces.

Recently experiments (47) have been carried out on reinforced concrete beams to determine the correlation between their flexural strengths and cube results on the one hand and pulse velocity tests on the other hand. This work has shown clearly that pulse velocity measurements provide a more precise estimate of beam strength than that given by cube test results. There are several possible reasons for this but undoubtedly the most important is that pulse tests are made on the actual material which is called upon to provide the strength for the beam rather than on supposedly representative samples as in cube testing.

Traditional testing procedure can hardly ever be very selective since it is impossible to know which part of a structure is represented by any particular cube particularly in view of the sort of variation shown in Fig. 6.3 which is for a beam made from a single batch of concrete. Some regions of a structure rely on concrete strength more than other regions do depending on how highly stressed these regions are and on the contribution which the steel reinforcement makes to the strength of the member. Nevertheless when test pieces are tested and they fail to comply with specified requirements, the structure could stand condemned whether or not the samples represent concrete in critical regions of the structure. Non-destructive testing can be much more selective and should allow the importance of the concrete in different parts of a structure to be properly taken into account.

Feedback of information from test results often allows subsequent work to be improved and this feedback can usually be more effective when the information comes from non-destructive tests because it is usually more comprehensive than that obtained from destructive tests on random samples, whether these samples are test specimens or units selected from precast concrete production. Some types of non-destructive tests provide useful information at a very early stage of production. For example gamma-ray backscatter tests can be used to measure density of local regions of precast units immediately after the concrete has been placed and, if local density has been correlated with the required performance of the units, the test results available within minutes can be early enough to allow effective remedial action to ensure the satisfactory performance of every unit made.

In conclusion, there is a strong case for a wider use of non-destructive tests for concrete inspection both for trouble-shooting and for routine quality control. The potentialities of these techniques can only be fully realized when results are treated realistically on their own merits even though this may require some abandonment of tradition-ally accepted definitions of concrete quality which are often only appropriate to destructive testing with all its limitations.

REFERENCES

1. FACAOARU, I. (1969). 'Non-destructive testing of concrete in Romainia', *Paper 4, Symposium on Non-destructive Testing of Concrete and Timber* (June) (London: Institution of Civil Engineers).
2. JONES, R. (1962). *Non-destructive testing of concrete* (Cambridge University Press).
3. BROWNE, L. J. I. (1968). 'Non-destructive testing of concrete—a survey', *Non-destructive testing* (Feb.), pp. 159–64.
4. JONES, R. (1969). 'A review of the non-destructive testing of concrete', *Paper 1, Symposium of Non-destructive Testing of Concrete and Timber* (June) (London: Institution of Civil Engineers).
5. BRITISH STANDARDS INSTITUTION (1969). *Non-destructive methods of test for concrete*, BS 4408 : 1969.
6. —— (1970). *Methods of testing concrete*, BS 1881 : 1970.
7. EVANS, E. P. (1960). *The effects of curing conditions on the physical properties of concrete*, PhD thesis, University of London.
8. BARON-HAY, J. K. (1960). *Structural properties of lightweight concretes*, PhD thesis, University of London.
9. SMITH, E. E., and WHIFFIN, A. C. (1952). 'Density measurements of concrete slabs using gamma radiation', *Engineer*, **194**, 278–81.
10. HARLAND, D. G. (1966). 'A radioactive method for measuring variations in density in concrete cores, cubes and beams', *Magazine of Concrete Research*, **18**, no. 55, 95–101.
11. BROCARD, J. (1955). 'The application of artificial radioactivity to non-destructive testing of materials', *Rev. Mat. Constr. et Trav. Pub. Nos. 482 and 483*, pp. 142–8.
12. ZHUKOR, V. S., KRYLOV, N. A., and SUDAKOV, V. V. (1958). *Dak. Mezvusovskoi Konf. po Ispytaniyam Soouzhenii*, Leningrad, pp. 205–14. Translation RTS 2163 National Lending Library for Science and Technology, October 1962.
13. POHL, E. (1962). 'Nuclear techniques in building', *Cement, Lime and Gravel* (May), 137–41.
14. HUYGHE, G., and MORTIER, P. (1955). 'Mesure de la densité du béton par absorption du rayonnement gamma', *Symposium National sur les Radioisotopes* (Bruxelles).
15. HASS, E. DE. (1953). 'Radioactive inspection of concrete', *Ontario Hydro Research News*, **5**, no. 4.
16. PREISS, K. (1965). 'Measuring concrete density by gamma ray transmission', *Materials Research and Standards*, **5**, (5), 285–91.
17. AMERICAN SOCIETY FOR TESTING MATERIALS. (1960). 'Symposium on nuclear methods for measuring soil density and moisture', *ASTM STP*, no. 293 (Philadelphia Pa.: ASTM).
18. SMITH, P. C., JOHNSON, A. I., FISCHER, C. P., and WOMARK, L. M. (1968). 'The use of nuclear meters in soils investigations. A summary of worldwide research and practice', ibid., no. 412 (Philadelphia Pa.: ASTM).
19. HUGHES, C. S., and ANDAY, M. C. (1967). 'Correlation and conference of portable nuclear density and moisture systems', *Highway Research Record*, **177**, 239–79.
20. IGNATIK, A. C. (1961). 'Use of radioactive isotopes to test compaction of concrete road pavements', *Jl Am. Concrete Inst. Proc.* **58**, no. 4.
21. RADZIKOWSKI, H. A. (1960). 'New techniques in highway construction', *American Road Builder* (Dec.).
22. DAVIN, M. (1964). 'Remarques sur des expériences de gamma densimétrie du béton', *Ann. Ponts et Chaussess (Paris)*, **134**, (6), 603–31.

23. PREISS, K., and NEWMAN, K. (1964). 'An improved technique for the measurement of density of concrete and soils', *Prov. IV Int. Conf. Non-destructive Testing, London, 1963*, pp. 135–41 (London: Butterworths).
24. FORRESTER, J. A. (1961). 'Über die Verwendung radioaktive Isotope im Bauwessen', *Wiss. Z. Hochsch. Bauw. Leipzig*, **7**, 37 (see ref. 13).
25. SIMPSON, J. W. (1968). 'A non-destructive method of measuring concrete density using backscattered gamma radiation', *Bldg Sci.* **3**, (1), 21–30.
26. PRICE, B. T., HORTON, C. C., and SPINNEY, K. T. (1957). *Radiation shielding* (London: Pergamon).
27. ROCKWELL, T. (Ed.) (1956). *Reactor shielding design manual* (New York: McGraw-Hill).
28. RASO, D. J. (1961). 'Transmission of scattered rays through concrete and iron slabs', *Health Physics*, vol. 5, pp. 126–41 (N. Ireland: Pergamon).
29. CHILTON, A. B. (1964). 'Backscatter for gamma rays from a point source near a concrete plane surface', *University of Illinois Bulletin*, **62**, no. 26 (Oct.).
30. KÜHN, S. H. (1963). 'Effect of type of material on nuclear density measurement', Nuclear measurements 1963 Highway Research Record No. 66. Highway Research Board of National Academy of Science National Research Council publication 1250.
31. LEWIS, W. A. (1965). 'Nuclear apparatus for density and moisture measurements—a study of factors affecting accuracy', *Roads and Road Construction* (Feb.), 37–43.
32. GARDNER, R. P., ROBERTS, V. C., HUGHES, C. S., and ANDAY, M. C. (1967). 'A calibration model for optimising the air gap method of compensating nuclear density gauges for soil composition variations', *Journal of Materials*, **2** (1), 3.
33. PAWLIW, J., and SPINKS, J. W. T. (1957). 'Neutron moisture meter for concrete', *Canadian Journal of Technology*, **34** (8),
34. WATERS, E. H., and MOSS, G. F. (1966). 'Estimation of moisture content by neutron scattering: effect of sample density and composition', *Nature*, **209** (5020), 287–9.
35. MULLINS, L., and PEARSON, H. M. (1949). 'The X-ray examination of concrete', *Civ. Engng Publ Wks Rev.* **44** (515), 256–8.
36. WHIFFEN, A. C. (1954). 'Locating steel reinforcing bars in a concrete slab', *The Engineer*, **197**, 887–8.
37. FORRESTER, J. A. (1958). 'Application of gamma radiography to concrete', ibid. **205**, 314–15.
38. —— (1959). 'The use of gamma radiography to detect faults in grouting', *Magazine of Concrete Research*, **11** (32), 93–6.
39. —— (1969). 'Gamma radiography of concrete', *Paper 2, Symposium on Non-destructive Testing of Concrete and Timber* (June) (London: Institution of Civil Engineers).
40. HONIG, A. (1960). 'Zjist ovani oceloro uyztuze jadernym zarenim', *Pozemni Star*, **11**, 579–86 (see also ref. 13).
41. HMSO (1969). Ionizing Radiations (Sealed Sources) Regulation 1969 No. 808 Factories.
42. LIEBERMAN, R., et al. (1961). 'The development of radioactive tracer quality control systems', Report No. BMI-1508, Battelle Memorial Institute (27 March). Prepared for US AEC, OID.
43. IDDINGS, F. A., et al. (1968). 'Nuclear techniques for cement determination', *Engineering Research Bulletin No. 95*. Division of Engineering Research, Louisiana State University.
44. KOLEK, J. (1958). 'An appreciation of the Schmidt rebound hammer', *Magazine of Concrete Research*, **10**, no. 28, 27–36.

45. —— (1969). 'Non-destructive testing of concrete by hardness methods', *Paper 3A, Symposium on Non-destructive Testing of Concrete and Timber* (June) (London: Institution of Civil Engineers).
46. LEVITT, M. (1969). 'Non-destructive testing of concrete by the initial surface absorption method', *Paper 3B*, ibid.
47. ELVERY, R. H., and DIN, N. M. (1969). 'Ultrasonic inspection of reinforced concrete flexural members', *Paper 5*, ibid.
48. HALSTEAD, P. E. (1951). *Technical Report No. TRA 197* (London: Cement and Concrete Association).
49. REIDING, F. J. 'A portable reinforcement cover meter', *TNO Report No. B1 61–25/103* (Holland: Inst. TNO voor Bouwmaterialen en Bouwconstructies Delft).
50. TOBIO, J. M. (1958). 'Empleo del bloque de yeso en la determinacion de lumedad en las pastas de cemento', *Revista ION publication No. 209* (Dec.).
51. BOOT, A. R., and WATSON, A. (1964). *ASTM STP*, no. 373 (Philadelphia Pa.: ASTM).

7
Building in hot climates

Balwant Singh Saini, PhD, BA, BArch, FRAIA,
FRIBA
Department of Architecture and Building,
University of Melbourne

7.1. Hot climates

7.2. Building design
7.2.1. Design for human comfort
7.2.2. Structure
7.2.3. Glass
7.2.4. Roofs
7.2.5. Floors
7.2.6. Design for extreme climatic conditions

7.3. Building materials
7.3.1. Materials in hot humid climates
7.3.2. Materials in hot dry climates
7.3.3. Structural materials
7.3.4. Paints and bituminous materials

7.4. Labour
7.4.1. Availability
7.4.2. Efficiency

7.5. Building techniques
7.5.1. Construction sectors
7.5.2. Industrialized building

7.6. Housing
7.6.1. The housing shortage
7.6.2. Self-help schemes

7.1. HOT CLIMATES

Building in hot climates often means building in tropical regions, but the use of this term could be misleading since, according to the *Shorter Oxford English Dictionary*, the tropics mainly refers to the hot and humid areas lying 'between (and around) each of the parallels of latitude distant about 23° 28′ North and South of the Equator'. It ignores the fact that large parts of the world, particularly deserts, which are equally hot, extend well beyond these geographic limits. Besides, the climatic characteristics of a region are normally assessed on the basis of the prevailing atmospheric conditions which include temperature, moisture, wind, and evaporation. Therefore, the geographic definition of the tropics indicates the prevailing climate to the extent that the temperature depends upon the incoming solar radiation, which in turn depends upon the latitude. It implies that the area experiences high temperatures with relatively small seasonal and diurnal variations. It fails to provide a basis for differentiating the several contrasting climatic types to be found within the boundaries of the area.

Various definitions have been advanced to meet the needs of specific research, and the criteria selected have brought about a variety of classification systems. Of these, three are best known. All use temperature as a common denominator, the other elements being precipitation (Koppen), moisture index (Thornthwaite), and vapour pressure (Lee). Koppen's system of climatic classification is simple and the most widely accepted (1). Based on the criteria of plant growth and distribution, it defines eleven climatic types with a code for every subdivision where necessary. As hot climates are demarcated by the 64·4°F (18°C) isotherms for the coldest month in each hemisphere (January and July) within this region Koppen recognized the existence of two main types: rainy climates, which include tropical rain forest, monsoon and savannah, and hot dry climates, which include steppe and desert types.

Buildings are primarily designed and constructed to provide shelter for humans, animals and goods. From the point of view of human

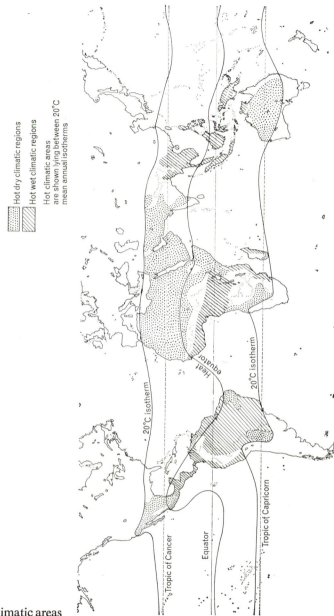

Fig. 7.1. Hot climatic areas

comfort, climatic regions could well be divided into fairly distinct zones, whose climates are based upon the classifications set out by Atkinson and others (2). According to this classification, hot climatic zones include the warm or hot humid, hot dry, upland, tropical island, savannah, and monsoon and maritime desert. Of these, the last four are comparatively insignificant in the global context as the bulk of hot climatic areas experience either hot humid or hot dry conditions (Fig. 7.1). Hot humid conditions, where the temperature remains between 80 °F (26·7 °C) and 90 °F (32·2 °C) and where humidity is extremely high, occupy much of equatorial Africa, the Amazon basin, and most of south-east Asia from the south Indian coasts to Papua and New Guinea. Hot dry climates where summer temperatures are high, winters are temperate and where the diurnal range is large, prevail in North Africa, the Middle East, coastal Peru, south-west America, and large parts of Australia.

In addition to a hot climate, the regions mentioned above possess many other things in common. With one or two minor exceptions, most of them are not only experiencing a rapid social change, but they are also underdeveloped economically and technologically. Compared to temperate areas, there are important differences in the organization of their building industry, its finance, and in the skills, training, and responsibilities of its members. Regulations, government procedures, and commercial practices are also different.

It is proposed here to examine broadly some of these aspects, with particular emphasis on problems of building design, materials, labour, and techniques. It must be realized, however, that in view of the extraordinarily large range of conditions which exist in hot climatic regions, the discussion must at best be general; conditions which exist in a particular place may not be typical of other areas. However, there are enough distinct areas which stand out significantly and which form the basis of this present analysis.

7.2. BUILDING DESIGN

7.2.1. Design for human comfort

Building in hot climates involves design and construction. An efficient but expensive way to achieve physical comfort in a building is to install a complete air-conditioning system. However, very few people can afford such a luxury. This is particularly so in areas remote from urban centres where air-conditioning systems are not only expensive to install but are very costly to operate. Invariably, there are extraordinary maintenance problems.

By and large, air-conditioning is generally confined to such buildings as operating theatres in hospitals, a factory for the manufacture of precision instruments, or other structures where utmost control of temperature and humidity is of great importance. For buildings which primarily accommodate people, air-conditioning is probably only used in luxury hotels which cater for those tourists who expect it and can afford the extra cost involved.

This situation has given a tremendous challenge to architects, engineers, and builders in developing countries who are faced with the responsibility of providing comfort in buildings by means other than air-conditioning and similar expensive systems of artificial climate control. The question is how to design buildings which provide comfortable indoor conditions through suitable selection of materials, construction techniques, planning, and orientation.

In order to plan and build for physical comfort, it is necessary not only to understand the nature of the climate and the reactions of people to it, but also its impact on building components and their performance under particular conditions. This means that factors such as temperature, humidity, rate of air movement and radiation from walls, floors, ceilings, and other surrounding surfaces, all of which contribute to human comfort, must be converted into a language which can be expressed in the design of building.

In hot humid climatic zones, the human body needs relief from heat as well as humidity. It finds it difficult to cool down by losing heat in the evaporative process. To aid evaporation from the body, the air movements assume a very important role. In the hot dry zones, for comfort conditions, reduction of heat takes precedence over the air movement. To achieve this it is necessary to create heavy shade internally and externally, a situation which, combined with measures to reduce intense glare and dust penetration, affects the design of buildings and their orientation. It is a situation which is the opposite of what is desired in hot humid areas where comfort requirements call for currents of moving air which, if supplied by prevailing breezes, demands suitable orientation of buildings with open walls for ventilation. Generally, it is far more comfortable to live in hot dry zones than in hot wet ones.

7.2.2. Structure

One of the chief causes of discomfort inside buildings in hot climatic regions, is high temperatures generated by heat from solar radiation. There is a considerable body of literature available to architects and builders which indicates ways and means of reducing this impact. Apart from orientation which involves the study of sun path determination and analysis of sunshading requirements which has led to the design of such shading devices as louvres, there is a need to look carefully at the function of the structure in controlling the thermal environment. Carefully selected materials are able to considerably reduce heat gains due to conduction and radiation from the surroundings. Conduction gains are minimized by insulation and a reduction of the external surface area, while radiation gains are reduced by shading, reflective coatings, non-reflective surroundings, and orientation.

The extent of heat gain through conduction can be reasonably controlled by the appropriate selection of materials, which is achieved by assessing the comparative performance of the various materials in terms of their thermal conductivity value or their over-all heat transmission coefficient, known as the k and U-values respectively. The assessment of heat transfer by radiation is difficult, as the amount of heat received on the several faces of the building varies continuously throughout the day, depending upon whether they are sunlit or not, and on the angle of incidence of the sun's rays as it moves through the sky.

The amount of heat transfer through the shell of a building not only depends upon the insulation value of the material, but also on the heat storage capacity. Dense or heavyweight materials such as mud, brick, or stone, have a very high heat storage capacity. They take a considerable time to heat up, and once heated, take a long time to cool down.

Materials with a high heat storage capacity have a great advantage in hot dry regions where they take a long time to absorb most of the heat received during the day before it is passed on to the inside surface. This behaviour is particularly admirable in such buildings as

Fig. 7.2. Building in hot, dry regions. Flats at Beersheva in the Negev Desert, Israel
Architect: Hanan Mertens

Fig. 7.3. Building in hot, humid areas. Secondary school building in Villa, New Hebrides, South-west Pacific
Architects: Warren and Mahoney of Christchurch ,New Zealand

schools and offices, where a cool interior is assured during the normal daytime occupancy. At night, when the outside temperature drops, the interiors are then too warm for comfort. To overcome this problem, people in traditional desert settlements, in North Africa and the Middle East for instance, move outdoors to live and sleep in courtyards, verandahs, or the roofs.

In a warm humid climate or where nights are only slightly cooler than the days, heavyweight construction is at a disadvantage since the cooling-off process at night is so slow that the indoor temperatures stay far too high for comfortable occupancy. In such areas, lightweight structures perform better as they cool down rapidly even when there is a slight drop in outside temperature. There are many parts of the world where hot climatic regions experience both wet and dry conditions. A possible solution here lies in combining light and heavyweight construction which should ensure cool living spaces in some part of the building at all times.

7.2.3. Glass

A source of heat gain, or loss for that matter, which is often overlooked by many designers, is caused by the provision of large areas of glass in tropical buildings. Most authorities on tropical architecture are unanimous in their view that glazed areas should be kept to a minimum compatible with adequate natural lighting.

Glass has little insulation value and heat will flow through it wherever there is a difference in temperature from one side to another. Weiss maintains that heat transmission through a single sheet of shaded glass is about sixteen times that through a shaded timber-framed wall insulated with 4 in of mineral wool (3).

In an example calculated for a simple structure in Pretoria, Richards found that during the afternoon of a summer day with unprotected windows in its 9 in brick walls, the ratio of the instantaneous heat gains through equivalent glass and wall areas is about 7 to 1 for the north wall, 3 to 1 for the east wall, 11 to 1 for the south wall, and 40 to 1 for the west wall (4).

Heat-absorbing glass can reduce the transmission of solar radiation but its effectiveness is limited because its own temperature is raised in the process, which in turn causes an increase in the heat convected and re-radiated into the room. Transmission is delayed rather than stopped, unless the glass is used independently of the structure itself and set away from the wall in a free-standing position. These observations seem to indicate that the use of glass in a building should be reduced to an absolute minimum, until such time as the glass industry is able to manufacture a product which reflects a higher percentage of heat. There are indications that new products such as Spectrafloat glass may in fact provide a realistic answer to the problem.

7.2.4. Roofs

Because of its orientation and comparatively large exposed area, the roof can be a major source of heat gain in a building. Intense radiation from the sun generates considerable heat during the day which, after transmission, raises the temperature of the ceiling surfaces. According to Richards, this radiation from heated ceilings not only heats other surfaces but indirectly adds to bodily discomfort (5). One way to minimize the transmission of heat is to reduce the proportion of radiation absorbed by the top surface. Although there is little reliable evidence which proves the value of roof insulation, the contribution of the reflective quality of the roof should not be underestimated. Richards, however, rightly points out that 'in regions of dusty atmospheres, the normal dust deposit collecting on top of a roof can seriously reduce the efficiency of reflective insulation and can reduce the effectiveness by half in one year' (6).

Most heat flow observations seem to indicate that the amount of protection afforded by different roofs of conventional construction and materials varies considerably. In a series of records made in the hot humid towns of Accra, Bombay, Colombo, Bangkok, Singapore, and Kuala Lumpur, Koenigsberger and Lynn have found that of the forty-two roof types investigated, less than half the examples satisfied the desirable conditions, namely that 'temperature of the underside of the ceiling should never be more than $8\,°F$ ($4\cdot4\,°C$) higher than the dry bulb temperature' (7). The study also found that unless repeatedly painted by lime or paint, both corrugated iron and asbestos sheets were found to be very poor roofing materials. Aluminium is better, but even this, as with other sheet or tile roofs without ceiling or special insulating layers, is not likely to provide adequate heat protection in the warm humid tropics.

Massive flat roofs are useful only for buildings such as offices or schools which are not used during the night. Concrete slabs introduce a time lag into the heat transfer which results in considerable delay before it cools down after sunset. The extent of this time lag depends upon the thickness of the slab. The time lag between the upper and lower surface temperatures for a 6 in (150 mm) concrete slab, for instance, is five hours. For an 8 in (200 mm) slab it is six hours. To increase this lag, such roofs require a 3 in (75 mm) thick sand and cement screed and a fibreboard ceiling.

Fry and Drew, who are well known for their buildings in West Africa and India, have advocated the use of a double roof system with an air space between (8). The reason offered is that the upper layer of the roof has a reflective surface which throws off a high proportion of the sun's heat. The hot air which might accumulate between the upper and lower levels is allowed to escape at the open ends, a process which is often helped by the air flow and by the prevailing wind if the building is orientated correctly. This system is used in a number of buildings in West Africa, particularly those in the University College, Ibadan.

Research reports, however, indicate that natural ventilation of roof ceiling spaces has no effect in the summer on the temperature of the air indoors. This view is based on a number of laboratory and field studies which include Givoni and Sheldon's work in the hot dry areas of Israel and Koenigsberger and Lynn's observations made in hot humid towns (9). Both studies came to the conclusion that heat was transferred from roof to ceiling, mainly by radiation. Convection played a subordinate role. Generally, most observers seem to agree with the view that thermal and reflective insulation—particularly in the ceiling—is by far the most effective means of reducing the lower surface of ceiling and room temperatures and that once achieved, natural ventilation of the roof–ceiling space is no longer important.

7.2.5. Floors

Floors can be constructed in timber or concrete. They may be on the ground, just above ground, or raised sufficiently high as is done in many buildings in hot, humid coastal areas where they virtually double the useful covered area. The usual reason given for high-level floors is that they facilitate the inspection of termite attack. It is difficult to justify this in view of the extra costs involved.

In hot arid areas, a concrete slab floor exerts a temperature stabilizing influence on indoor temperatures. It is also maintained that in such areas the floors, if suspended, should be enclosed against the entry of hot winds which would heat up the floor itself. For hot humid areas, however, the thermal stabilizing factor is of little importance and so it is inconsequential what kind of floor is provided in a building.

Although there is an increasing realization about the importance of right selection of materials appropriate for specific climatic requirements where they are built in sympathy with their surroundings, there are many who ignore the simple rules. Instead of following closely the sensible approach used by indigenous people of hot climatic regions, they often employ materials and techniques foreign to the region, and are therefore unsuitable for the purpose for which they are constructed.

7.2.6. Design for extreme climatic conditions

Building methods are as much affected by the type and performance of materials as by structural considerations. The latter present the same problems with buildings in hot climates as elsewhere, but there are some which occur with such extraordinary severity as to require special measures, in terms of design and construction. Among these problems, the most important requires structures to resist *earthquakes, cyclonic winds* and to resist the forces exerted by *plastic soils*.

Earthquakes consist of vertical and horizontal ground vibration. The horizontal motion is generally greater and this requires a structure that will offer resistance to lateral forces. Economic earthquake resistance design for tall structures requires a knowledge of mechanics beyond the scope of this article. However, the design of a small rigid building of one or two storeys in height is not so complicated a problem and simple rules have been devised and can safely be used. These are still invariably misunderstood by untrained builders in some developing countries who do not have the benefit of technical advice.

A frequent fault commonly noticed by the writer is the omission of a so-called tie-beam, on top of the walls. If provided, it is rarely carried right around the perimeter of the building in order to tie the whole structure together. Tie beams at footing level are also essential if the building is founded on piles or on ground worth less than 2000 lbf/ft² (96 kN/m²) allowable bearing pressure.

From the design point of view, it is important that the centre of gravity of a building and the centre of the elements making for stiffness should be as close as possible so that the building will not be subjected to a twisting effect during the earthquake. A symmetrical disposition of cross-walls and stiff partitions thus offers the best possible condition for rigidity and stiffness. Open-sited buildings with large open floor spaces which attempt to carry shear forces on one or two cross-walls are not good in an earthquake area. This is particularly so when the stiffened walls are collected towards one end or another of the floor plan. These limitations are of particular significance in hot humid climatic regions where buildings are required to be open and spread out for aiding cross-ventilation.

Many hot climatic areas, particularly the islands and some coastal regions, experience quite severe winds which create major structural problems. The most extreme form of damage caused by the impact of cyclonic or other strong winds on buildings results in simple structural collapse of joinery frame, doors, or the structure itself. The roof is invariably lifted off the structure through extreme wind pressures or leeward suction on the building. The strong winds also tend to exploit any area of weakness in the structure, be it loose sheets of iron, guttering, or shutters and doors. The continual hammering and vibration results in a failure of these sections, which are ripped open and blown away, thus starting a possible train of collapse of the rest of the building. Flying debris is another major source of damage.

Codes for Minimum Design Loads in Buildings, specify clearly the wind loads that buildings in particular areas should be designed to resist. Publications of Building Research Stations offer valuable directions for designers and suggest some sound procedures for structural details. Among these are included the need to provide suitable anchorage to tie securely together all elements of a building, thus preventing lateral movement and the overturning of a building as a whole. Concrete slab roofs are heavy and strong enough, but roof tiles need to be secured individually. Corrugated iron sheet roofing, a popular material in many hot climatic regions, must be screw-fixed rather than nailed, and special attention may have to be given to a close spacing of fixing clips of strip-metal roofing and to the securing of the clips to the roof framing. Particular attention needs to be paid to glassed areas, shutters, window, door hinges, and other fittings.

Plastic or expansive soils are found in many hot climatic regions. They present quite severe problems to building foundations. In order to be reasonably free from disturbance by moisture changes, footings in some areas are required to be as much as 15 ft (4·6 m) below the surface. Loads acceptable as safe bearing values differ according to the nature of the plastic soils. They range from 0·5 tonf/ft² (54 kN/m²) for quicksand and alluvial soils, to 2 tonf/ft² (215 kN/m²) for ordinary clays and 4 tonf/ft² (429 kN/m²) for hard clays. Because of the complex soil phenomena associated with small or domestic construction, these values have to be considerably increased. A large range of designs for footings have been evolved to suit various situations.

The subject is indeed vast and complex and beyond the scope of the present discussion. Most of the measures suggested by research organisations point to construction techniques which require additional expense compared to that necessary for structures in temperate climatic areas.

7.3. BUILDING MATERIALS

In most hot climatic regions, building materials and construction techniques used for low cost building programs are directly influenced by climatic, social and economic conditions prevailing in the region.

7.3.1. Materials in hot humid climates

In hot humid areas, owing to high rainfall, the traditional materials are organic, such as timber, grasses, reeds, leaves, canes, and bamboo. Of these, only timber has proved durable and sturdy enough to be acceptable as a building material under modern conditions. The others, though impermanent, still continue to be used in rural areas and where the pressures of population have not depleted the supplies.

The structures made from them are light and can, therefore, withstand earthquakes and other shocks. They are cool and their walls are ideal for cross-ventilation in the humid tropics where this is of utmost importance. Fabrication and later repairs are easy and their reclamation value is fairly high. The materials have an obvious role

Fig. 7.4. In hot, humid areas, owing to high rainfall, the traditional materials are organic, such as timber, grasses, reeds, leaves, canes, and bamboo

to play within the under-developed economy of many countries and it is not advisable to dismiss their potential.

Few of these materials, however, are able to survive the numerous agents of decay, such as weather and insects. In some regions, the average life of thatch roofs and wall mats rarely exceeds four to five years. The fire hazard is usually great and as a consequence, there have been incidents when whole villages have been reduced to ashes.

There are ways and means to improve their performance and reduce deterioration due to weathering and insect attack by the use of preservatives and other chemical means (10). Preservatives such as pentachlorophenol in oil are used outside where they are exposed to weather but their disadvantage lies in persistent odour and an oily surface which is impenetrable. Fixed water-borne salts are all copper/chrome/arsenic complexes which, if used in the right concentration, provide excellent protection against decay, borers, and termite attack. Unfixed salts are much cheaper and easily penetrate into loose fibrous materials. On the other hand they leach out just as easily, especially when used outside or in contact with the soil. Simple organic salts, such as the boron type, are used for the control of borer attack.

Preservation processes range from simple brush coating and dipping treatments through sap replacement, open tank or convection processes, to pressure treatment. The selection depends largely upon the circumstances of use, kind of material to be treated, and the capital and time available.

Perhaps the best protection against fire is to isolate all possible sources of heat, but fire-retardant coatings are of considerable value. They are generally made from boric acid or ammonium phosphate. The effectiveness of preservatives in high rainfall areas is fairly limited since they have a tendency to leach out. Here it may be possible to press some of the low cost grasses, leaves, and reeds into simple building boards (11). Under primitive conditions, these boards offer many advantages. There is considerable saving in time and labour costs as they are large structural units, light in weight and compact in nature. Transport is also easier, which makes them very adaptable for prefabrication. Manufacture of these boards, of course, necessitates the setting up of local cottage industries and workshops where unskilled labour could be trained to supply the limited needs of the rural building industry.

7.3.2. Materials in hot dry climates

In hot dry regions, earth in one form or another constitutes the basic construction material for low cost building. Traditionally, communities in North Africa and the Middle East have used this material where rainfall is low, timber is scarce, and the climate dictates the use of heavy materials with a high heat storage capacity.

Earth construction has mainly been evolved in four forms (12). The oldest is wattle-and-daub-and-cob, which is basically mud plastered over split bamboo or timber reinforcement. The later type are pise and adobe. Wattle-and-daub-and-cob are primitive means of construction which continue to be used extensively in rural areas. Pise and adobe lend themselves to the conventional masonry method of building.

The term 'adobe' refers to mud or puddled earth and is applied to the building of sun-dried bricks. The term 'pise' is used for rammed earth, which is laid in damp or moist form between temporary moveable formwork. Each construction has generally high manual labour content. Large quantities of earth are necessary for thick walls as it is weak in lateral stress. This, no doubt, offsets whatever advantages there may be. It is a slow process and much time is taken in waiting for the earth to dry out. Use of this method is only feasible in areas where transport is not available to offer an alternative material; or where it is largely built on a self-help basis by the personal efforts of the owner; it is an effort which is rarely measured in monetary terms.

A major disadvantage of earth construction is that the method does not lend itself to smooth and clean finishes, particularly in monsoonal areas, where it tends to crumble and needs continuous maintenance after each rainy season. Although some waterproofing is achieved by oil coatings and some rendering, the maintenance is still time-consuming.

Over the last two decades, there has been a widespread use of stabilizing agents such as bitumen, lime, and/or Portland cement, all of which help to bind the soil and provide it with added strength and durability. Soil stabilization in simple terms means 'restoring, and in some cases, improving, the original properties it had before being removed from its natural form' (13). Stabilization is achieved by subjecting the soil to high pressure by manual or mechanized means after mixing in a cement substance.

To facilitate this process, soil is moistened up to its optimum grade; i.e., up to the point where it is wet enough to allow maximum compaction. The addition of cement helps to bind the sand particles to the clay content of the soil. The amount of cement necessary for stabilization largely depends upon the quality and proportion of clay. Soils with a high clay content require a high cement ratio. Clays of appreciable shrinkage quality also need more cement than clays of more stable properties. Depending upon these characteristics, the proportion of cement can vary from 4 per cent to as much as 16 per cent. The ratio is critical as on it hinges the economy and advisability of utilising soil cement products in building projects.

7.3.3. Structural materials

Among the manufactured materials which have so far made the greatest impact on low cost building in developing countries are iron, steel, and cement. The building products which result from them and are widely used, are corrugated iron, concrete blocks, and reinforced concrete.

These materials have begun to make inroads into the area where traditional materials have always been accepted. Perhaps the first non-traditional material which obtained widespread acceptance was corrugated iron. It has helped change the roofscape of many settlements in hot humid regions where the thatched roofs have given way to metal decking. It is easily transportable and simple to fix on site requiring a minimum of skilled labour. It presents obvious problems of corrosion, particularly near the sea, but in more protected inland areas, it is known to last as long as twenty years (14). An important advantage, despite its low insulation value is that it is fireproof and able to catch rainwater for drinking purposes.

During recent years, aluminium products have begun to make inroads into the market where corrugated iron prevailed; even though slightly more expensive, they have distinct advantages in terms of thermal comfort, appearance, and durability.

Lately, despite very many disadvantages, asbestos cement products have begun to replace galvanized iron. In areas where building materials have to travel long distances, the breakage rate of asbestos sheets is considerably high. In the cyclonic or hurricane areas, they are easily punctured by flying objects and therefore cause high winds to build up pressure within the building. In hot humid areas, algae growth on damp sheets, after a few years of exposure, leads to considerable blackening. This is not only unsightly, but reduces the effectiveness of the sheet as a reflector of solar radiation. In a report on 'The weathering and durability of building materials under tropical conditions', Holmes recommends the use of weak copper sulphate solution washed over the affected area (15).

In countries where the art of masonry construction is traditional, water is mainly responsible for damage. Earth walls from adobe or sun-dried bricks are particularly vulnerable. To avoid the erosive action of driving rain, in addition to provision of deep roof overhangs it becomes necessary to carry these materials on concrete or stone to a higher level, clear of storm water. In most areas, burnt bricks continue to be used in most buildings. They are durable, but sometimes the cost of manufacture can be high where fuel for kilns is expensive.

In some countries, there are moves to replace burnt bricks by concrete blocks which are cheaper and equally effective (17). Apart from stabilized soil products mentioned earlier which are a distinct improvement on the traditional mud walls, concrete blocks with a fairly high cement/sand ratio offer an excellent material for good-quality building. Their cost is still far too great to make a general impact on low-cost construction, though this may change with increased economic development and improved standards of living in poorer countries. In the meantime, a good compromise offered by the stabilized soil blocks which do not contain a high percentage of cement is likely to continue in use in preference to other masonry products. It is now possible to install reasonably priced hand-operated machines which can turn out anything from 500 to 1000 blocks per day, by employing no more than two unskilled workers.

Apart from water, which presents on great problem in hot arid areas, deterioration is also caused by the abrasive action of dust and sand under windy conditions. In a paper on tropical paints, Footner and Murray referred to the gradual abrasion of steel by successive dust storms (18). Brooks reports that stripping of permeable finishes and paintwork, generally in dust storms, is common and that, in Iran, the summer winds have caused slow erosion and undermining of walls and buildings (19). Dust is also a great menace to all kinds of mechanical equipment and fixtures. Control of dust and sand presents a number of complex problems which are impossible to detail here.

Deterioration by radiation is mainly caused either by photochemical action (which is partly due to slightly enhanced ultra-violet band in sunlight in low latitudes) or through the heat effect. In the case of the latter, expansion in the daytime is followed by contraction during the cool evenings, which leads to cracking and disintegration. The effect can be minimized by reflective treatment, by shading, or by the careful design of details of the components.

In hot humid areas, the high volume and high intensity of rainfall causes problems of water penetration and flooding. Prolonged wetting delays the drying of materials and this promotes fungal growth. Condensation caused by the daily fluctuation of temperature deposits moisture, often in confined spaces, from where it is difficult to ensure its evaporation. Most surfaces, when exposed to sunlight, encourage rapid physical and chemical breakdown. This problem highlights the need for shedding water and the provision of adequate ventilation and good drainage.

Attack by wood-destroying fungi is potentially far more serious than surface moulds, as they invade the cell structure of the wood, causing complete destruction. Of the wood-boring insects, termites are the most destructive (20). There are many different species of termites. The so-called 'subterranean termites' occur in many parts of hot climatic regions, but the 'dry wood termite' which do not require contact with the ground, infest timber and furniture. There is little that can be done to prevent their attack other than to use timber that is naturally resistant or which has been impregnated with a suitable preservative.

Subterranean termites can be controlled by a number of precautionary measures which include ant-caps or shields over stumps and piers suspended over frames, and by providing clear and free air circulation under floors to avoid dampness. Treated soil barriers (poisons) have also proved useful in some cases. There are methods by which concrete can also be made termite-proof. Some authorities recommend the addition of 0·5 per cent emulsion of Dieldrin to water when mixed with concrete.

Concrete is widely used in hot countries and in most cases is subject to little deterioration. The chief problem lies in getting the ingredients, particularly cement, which in some areas is imported at an exorbitant cost. Sand is available on riversides and creek beds and the same applies to aggregate, which may have to be transported over long distances. Stone suitable for crushing is only available in certain areas and even then the machinery for crushing it is only available in places where large engineering works exist.

But the most important need is the availability of reasonably clean water. Salty water, water with a phosphate content and water containing organic matter are considered unsuitable for making concrete. Water is also necessary for continual wetting of the surface of concrete work for at least two weeks. This is particularly so in hot dry lands where rapid evaporation and dehydration is known to cause low strength, cracking, crushing and excessive permeability.

In hot dry areas, usually when high winds are present, cracks in new concrete may be of two types; crazing cracks and plastic shrinkage cracks. Crazing cracks are caused by more rapid drying and shrinkage on the surface than in the lower part of the slab and are due to inadequate curing. Plastic shrinkage cracks are deeper and more widely spaced; they usually develop during finishing and are caused by the rapid loss of mix water to the subgrade, by dry aggregate or by the air, because of hot sun, extremely dry, or windy conditions.

Palmer W. Roberts, who is a captain in the US Navy and heads the Construction Division Committee of Adverse Weather, has admirably summarized the following main precautions necessary for concrete work in hot dry climates (21).

a. All concrete should have, ideally, a temperature of 73 °F (22·8 °C) at the time it is deposited in the forms. The temperature should never be lower than 50 °F (10 °C) and never above 90 °F (32·2 °C).

b. The temperature of fresh concrete is controlled by the temperature of the material and the mixing conditions. Mixing periods should be held to a minimum to reduce the build-up of heat in the mixing drum. If the temperature of the cement exceeds 170 °F (76·7 °C), care must be taken to avoid 'flash setting' of the cement. Direct contact between water and hot cement must be prevented.

c. Aggregates both coarse and fine should be stored where they are protected from excessive heat.

d. Cool mixing water is essential.

e. Admixtures and cement substitutes should be used only on the approval of the engineer or architect. They should not be used without technical advice and then preferably only after testing.

f. All forms or surfaces that are to receive the concrete, such as subgrade and reinforcing steel, should be protected against excessive heat and air currents. Wetting down around the work will cool the air and increase the humidity, thus reducing temperatures and evaporation from the concrete.

g. Work should be scheduled carefully so that adequate skilled personnel are available to handle and place the concrete rapidly immediately upon delivery.

h. Cold joints should be avoided.

i. In extremely hot weather new concrete should be shaded or covered with wet material. Curing operations should be started as soon as the concrete has set sufficiently to avoid surface damage, and it should be sprinkled systematically with cool water.

j. Continuous water curing should be maintained, starting immediately after finishing, for a minimum period of twenty-four hours on formed or unformed concrete. Loss of water from the concrete may be prevented by using sprayed-on white pigment, curing compounds or plastic membranes. Curing for at least seven days is necessary where high strength or a durable wearing surface is important.

k. During and after the specified curing period, every effort should be made to reduce the rate of drying by avoiding air circulation.

7.3.4. Paints and bituminous materials

Materials often require paint on exposed surfaces to prolong their performance. The latter tends to break down more frequently in hot climates. This deterioration is probably caused by high radiation which intensifies physical, chemical, and photochemical deterioration (16). In a hot humid climate, the performance of paints is further affected by frequent alternation of rain and sunshine and continuous high humidity. Despite claims by some manufacturers, the performance of most paints is poor in the tropical regions. An average-sized building, such as a house or a shop, constructed from a lightweight structure, would require painting every three to four years.

Bituminous materials, particularly in the form of roofing felt, were introduced in many buildings about 1946. It was soon evident that the life of these materials was much shorter in hot climates than in the cooler regions. Intense solar radiation is known to promote chemical reactions which result in 'embrittlement, cracking and crumbling'. Provision of highly reflective surface treatment has been suggested as a solution to the problem, but the need for continual maintenance has successfully eliminated the use of bituminous materials in warm climates.

7.4. LABOUR

7.4.1. Availability

Buildings are generally the result of production, assembly, and maintenance of materials and components involving a large labour force and a variety of skills. The availability of labour, its quantity, quality, and the techniques used, decide to a large extent the methods of building. They also affect the quality as well as the cost of building.

Since building is a labour-intensive industry, the availability of labour is a prerequisite to any programme. In many poorer countries, labour is abundant, cheap, and mostly unskilled. Efficiency and productive capacity are low. In some hot climatic regions such as Australia and the south-west of the USA, labour is scarce in relation to expanding industries. Owing to the high cost of living, it tends to be expensive, but this drawback is somewhat compensated by relatively high technical skills and the availability of cheaper mechanical equipment.

In the sparsely populated hot dry regions of Africa, the Middle East, and Australia, skilled building labour tends to be highly transitional and floating in nature. In some cases it is confined to areas where a constant building demand exists. For projects in remote locations, labour has to be imported from relatively few, widely scattered, main centres of population. Owing to insufficient demand and the inability of the small contractors to meet the high overheads involved in remote areas, it is not even possible to compensate the labour deficiency by increased use of machinery.

In order to attract highly skilled labour to remote areas, the employers are forced to pay extra wages for travel rewards and allowances. Overtime work with increased rates of salary has to be ensured so that spare time can be effectively utilized in a strange and lonely environment. Add to this the high cost of living and scarcity of accommodation and facilities, and the whole problem assumes uneconomic proportions. In a survey of building and planning in central Australia, the writer found that the cost of labour was often 100 per cent more than that found in the capital towns. The scarcity of building labour as a whole, and skilled labour in particular, in these areas presents a serious handicap to increased reasonably priced building activity.

Fig. 7.5. In poorer countries, labour is abundant, cheap, and mostly unskilled. Its efficiency and productivity is low
Photo: Administration of Papua New Guinea, Department of Information and Extension Services

7.4.2. Efficiency

Poor performance is often attributed to hot climates under which building workers are forced to operate. Most builders tend to agree with the view that workmen in hot climatic regions are generally capable of the same output per hour as their opposite numbers in the temperate regions. If output falls below accepted standards, the blame should be laid squarely upon the management. As one construction company manager once stated: 'Poor performance indicates one of two things, either inadequate supervision and direction or faulty organisation of the supply of materials'.

Macpherson, Lambert, and others do not share this optimism of the contractors to the same degree (22). They maintain that in hot regions, high intensities of radiation on exposed building sites considerably lower the efficiency of the workers. There are some special tasks; plumbing, painting on the roof, carpentry, working unprotected from the sun, laying drain-pipes in the trenches, or electrical work in the ceiling for example, where conditions become so unbearable that frequent rest pauses become absolutely essential. With the rise in body temperature, some reduction in mental and physical co-ordination also take place. A systematic study of accidents in the oil industry of the Sahara indicated that in plants employing an equal number of staff and having identical productivity, there is a striking parallel between the monthly graph of the accident rate and the rise in temperature.

Lambert also found that the deterioration of the productivity rate is a function of effort. The more skilled the operatives, the less is the deterioration. Lambert's records indicated a 25 per cent reduction in efficiency when workers operated under hot climatic conditions of 57 °C combined with 30 per cent relative humidity (23).

Many authorities have made suggestions to reduce the adverse effects of heat. There has been a proposal that if a roof was constructed first, then it should be possible for men to work under shade. The idea no doubt possesses some merit, but it is not always practical in cases where the roof is supported on load-bearing walls. Besides, there is little one can do about heat from the sun when work is to be performed on top of the roof or outside the walls.

The suggestions for reorganization of working hours is worth serious consideration in areas where temperatures are lower early in the morning or late in the afternoon. Work output could be improved if the hours are set to avoid the most uncomfortable periods. It could be achieved by either beginning the work earlier and finishing by mid-day or by dividing the hours into two phases and enjoying a long break in the middle and afternoon period. The general attitudes should be to work with, and not against, hot climatic conditions and this could be extended to such things as transportation, and storage and utilization of materials. For instance, metal tools used by local people, and which have been developed over generations of use, prove much better than those imported from elsewhere.

Difficulties associated with skilled labour in hot climatic areas have contributed a great deal to high costs of quality building. There is an obvious case for increased mechanization of the building industry in many areas so as to compensate for the deficiencies described earlier. Apart from fears of temporary unemployment, arguments against mechanization include the need for importation of machinery involving hard-earned foreign exchange and the necessity to launch a large-scale operation for training people in skills not at present available. The ability to import machinery will no doubt depend upon the level of foreign currency available, but the mechanization need not necessarily mean displacing men by expensive machines imported from other developed countries. It can also mean rationalization of building materials' production, distribution, and utilization in such a way that the maximum benefits are obtained in each situation.

In a report on 'Building in Rajasthan', Charles Cockburn suggests two ways to mechanise building in developing countries (24). The first method is on-site, where simple mechanical aids can be provided to manual labour. These include such items as small concrete mixers, simple weight batchers, tripods, and pulleys. The second method involves off-site operations relating to building components, such as concrete blocks, roof slabs, window and door frames. Factory employment in the manufacture of components might compensate for loss of site work, and the unemployment temporarily generated by mechanization could, in the long run, be overcome by increased building activity generated by cheaper construction.

In many developing countries it is realized that a training programme for building skills is more likely to succeed if it is incorporated as part and parcel of technical education. In fact it is a recognized objective of most governments responsible for the welfare and development of their people. All that is needed is the realization of its importance and to ensure that the teaching is based on the actual needs in the field. A definite drawback of machines is that they are less versatile than the labour that they replace. Whereas a gang of men levelling the ground for foundations may be called away to move other materials or unload a waiting vehicle, machines can only be employed for certain specialized repetitive jobs.

7.5. BUILDING TECHNIQUES

7.5.1. Construction sectors

The whole question of mechanization versus labour and materials sets the limit to the sophistication of building techniques. In an excellent appraisal of the economic significance of construction, Duccio Turin sums up the important sectors of construction in developing countries into Traditional, National Conventional, National Modern, and International Modern (25).

The Traditional sector has been estimated to be between 20 and 30 per cent of the total building activity. Most of it exists in rural areas, but there is a substantial amount of work carried out in and around rapidly expanding urban centres. The characteristic of the National Conventional sector is a mixture of traditional materials and techniques, with a few modern and industrially produced materials and components. Turin has listed corrugated iron sheets for roofing, cement blocks for walls, simple reinforced concrete beams and lintels, crude joineries, and glazing as some of them.

The so-called National Modern sector involves the use of local technical and managerial skills in handling modern technologies, imported or adapted from more industrialized countries. Most of the building in this sector is carried out in capital cities and major urban centres. The International Modern sector draws on the most advanced construction technology available in the world. It is controlled largely by foreign consultants and often acts as a catalyst to the development of modern techniques. A prerequisite for its effectiveness is the existence of the National Modern sector mentioned earlier.

7.5.2. Industrialized building

In many under-developed areas, where local building materials and labour are virtually non-existent and where specialized skills, machinery, services, and stocks are not available, industrialized building is often suggested as an answer to difficult local building problems. Prefabrication is the magic word used by many, but unless the situation is critically studied, the whole effort towards partially or fully industrialized building is likely to lose direction in the welter of possible alternatives with a consequent dissipation of effort.

The experience of governments in many developing countries has not been very happy. In Singapore, where construction companies who claimed that their construction systems could produce dwelling units more efficiently than traditional reinforced concrete frame production used by the Singapore Housing Board, became bankrupt during the course of the building contract.

There are many conditions prevailing in the developing areas which represent serious obstacles to the application of modern industrial techniques. Some of these are, the existence of rural population, low average density of population in some areas, great distances separating large and small urban centres, inadequate transport networks and facilities, and insufficient knowledge of the potential of local resources. Often, the market capacity is small and scattered. Under these circumstances, it is very difficult to justify the high capital investment necessary for setting up manufacturing plants for building components. Considerable improvement can, of course, be achieved by merely increased standardization and more economic use of human effort, materials, and time.

In the writer's experience in the hot arid region of central Australia, delivery of materials and components to the remote sites involves packaging, storing, and transport by rail, road, and ship, all of which tends to add to the freight and handling charges (26). Often there is delay in securing space on ships; and the limited and scattered demand makes it uneconomic to stock-pile materials at any specific location.

The only alternative for a contractor is to order materials from the metropolitan centres as and when required. Dealers charge extra costs for handling at the source which involves marketing, crating, and packaging in the warehouses of suppliers. Joinery, for example, has to be cut, assembled, and marked and finally dismantled for painting. Generally, cost increases caused by the damage of materials transported overland and at warehouses present a major problem; for not only the material often becomes useless, but also there is considerable delay in time and consequent wastage of man-hours on site.

In central Australia, a system of partial assembly on site has generally proved successful. Here, raw materials and components such as windows and doors are supplied directly to the site for partial manufacture and assembly.

Victor Jennings has suggested that the first and most obvious step

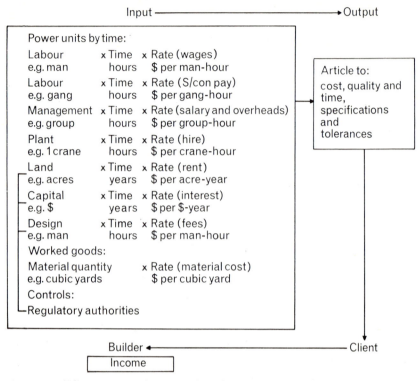

Fig. 7.6. Building resources in remote locations
After V. Jennings

in deciding a suitable course of action is to ascertain the various sources available and those needed to carry out the development of a building in a particular location (27). He suggested a systematic method of listing the various items under labour, management, plant, land, capital, design, controls, and worked goods; all of which have been assessed to suit certain specific requirements of cost, quality, and time (Fig. 7.6).

In the final analysis, the cost of industrialized building in most developing countries is likely to remain high. The industry as it is organized in most of these areas is not able to adapt to the changing requirements, and unless supply and demand is stabilized and rationalized, it is difficult to expect increased integration within it.

7.6. HOUSING

7.6.1. The housing shortage

One of the biggest challenges in the developing countries, of course, lies in housing which must be built at a low cost. At present, the standards are low and, what is more, are deteriorating rapidly by such fundamental factors as high rates of population growth, migrational floods of people from rural to urban areas, and low levels of economic development. The appalling conditions are epitomised by such large cities as Calcutta and Bombay and Djakarta, Manila, Saigon, Rio de Janeiro, Mexico City, and a dozen other impoverished agglomerations, bursting at their seams.

Calcutta, with its nearly 7 million inhabitants and growing by another $\frac{1}{2}$ million every year, already has 70 per cent of its families living in one room and uncounted thousands sleeping, starving, and dying on the pavements. The city is on the verge of a complete breakdown. Djakarta's swollen population of 3·5 million is augmented every year, not only by natural increase, but also by an influx of 100 000 newcomers looking for marginal employment.

The dimensions of the problem on a world scale are evident from United Nations statistics which indicate that dwellers in big cities in 1920 totalled 96 million, of which 80 million lived in temperate areas of the so-called western world, and only 16 million in developing countries; but by 1960, big-city dwellers numbered 351 million, of whom 139 million were in the poorer countries in hot climatic regions. In other words, while the metropolitan population of affluent countries had little more than doubled, that of the developing countries had multiplied five-fold.

In developing economies, where jobs and food take precedence over shelter, the housing shortage is likely to continue and the problem will become more severe, with all its adverse consequences. Apart from the high birth-rate, a major cause of the swelling populations of urban centres is migratory workers from rural areas who converge in the hope of employment and a better life. In the absence of any kind of accommodation, they build temporary structures with such materials as packing-cases, flattened tins, and corrugated iron sheets, often salvaged from factory waste and town refuse dumps.

They set up what are known as squatter settlements where proper services do not exist and water is often carried over long distances in tins or drums from public standpipes or private sources. Squatters are what A. E. S. Alcock called 'the greatest exponents of spontaneous unorganised self-help in the housing field, beginning by helping themselves to someone's land' (28). Under the present acute conditions, where housing finance is almost impossible to obtain and where there is no organized effort to provide for the housing needs of the people on a mass scale, self-help seems to provide the only alternative.

7.6.2. Self-help schemes

The idea of volunteering to join the community efforts for carrying out a particular task is not new to the people in hot climatic regions, where an attitude of mutual assistance has always been present, particularly in rural areas where communities, by the nature of their farming, are obliged to work together for purposes such as irrigation and harvesting. Gotong-rojong in Indonesia and Bayanihan in the Philippines are two such traditional demonstrations of co-operative effort.

In all self-help methods, the occupants contribute a reasonable share towards the cost of the house building (29). This can take the form of either money, materials, or labour. The term 'aided self-help housing', was perhaps used for the first time in some of the housing projects initiated in Puerto Rico in the early 1940s, though as early as 1936, self-help methods were used by a housing co-operative in Nova Scotia, Canada. Since the early 1950s the term has been widely used by various international organizations.

In 1953, the Inter-American Housing Centre in Bagota, Columbia, published a manual for the organization of a pilot-aided self-help housing project. This was perhaps the first real attempt to systematize the idea of self-help into a comprehensive method. Since then, a number of self-help housing programmes were launched in many countries with varying degrees of success. The first waves of enthusiasm were replaced quickly by a slow realization of many pitfalls and problems that accompany such ventures.

The chief attraction of co-operative self-help housing lies in the fact that many people in low-income groups are able to solve their housing problems even when the means at their disposal are limited. To achieve this, ways are found to reduce the cash contribution of individuals.

Housing co-operatives are generally organized to obtain sufficient housing credits for their members who pool down their limited economic resources in order to provide consolidated responsibility to the agency financing the loan. The cash contribution required by the co-operative can be reduced if members are able to supply voluntary labour for construction of the houses. Self-help is therefore generally understood as 'the voluntary participation of people in solving their own problems with their own resources and skills'. The word 'aided' implies some sort of external assistance which may be financial, technical, or both.

In the case of housing, aided self-help can take two forms. The first and simplest is that in which an individual and his family undertake the construction of their house. It may be called *individual-aided self-help* (IASH).

The second type is known as *group-aided self-help* (GASH) where a number of people pool their efforts, mostly manual, for the construction of their houses and eventually other amenities required by the whole community. Although more difficult to carry out, GASH projects tend to achieve far better social and educational results. They are ideally suited to the needs of co-operative societies as they imply group effort in the achievement of a common endeavour. IASH, on the other hand, makes little contribution to the development of a community spirit. In a development programme, where provision for housing is only a part of a larger concept, GASH is the obvious answer. However, any decision to adopt this system can only be made after a careful analysis of local conditions and problems.

During the last twenty-five years, many countries have attempted to utilize people's voluntary participation in the construction of their houses. Experiences of many large GASH housing projects have, however, focused attention on a number of difficulties. Most important of these include problems of building inefficiency which result from the lack of skills among participants and exaggerated periods of time required to complete the work. Both these shortcomings no doubt considerably contribute to the increased costs. But it must also be accepted that a great deal of education and strengthening of co-operative attitudes can be achieved through GASH methods.

Efficiency of the building operation is certainly impaired to some degree by the technical shortcomings of the participants. On the other hand, to take a more positive view, GASH could actually help to develop a nucleus of skilled tradesmen and thus play an important role in building up the manpower required in a developing country. According to one report from British Honduras, mutual self-help projects have considerably helped to increase the number of carpenters and masons in the region.

It is possible that in some areas, labour is so cheap and abundant that GASH projects may not appear to be worth the effort. In such cases, it may be better to provide remunerative employment for the spare time of participants so as to enable them to afford the cost of labour required by the project. However, a true assessment is difficult because such an approach ignores the educational benefits inherent in most GASH programmes.

The problem of unskilled participants can partially be overcome by instituting a training programme in building methods, before the start of the project. It has also been found advisable to:

  a. Hire a limited number of skilled labourers as foremen and trainers of participants;
  b. Complete beforehand one house of the project so that a complete model is provided for the rest of the work and difficulties are spotted in advance;
  c. Prefabricate, wherever possible, some of the components of the house so that the building operation is reduced to one of assembly, thus eliminating the risk of construction of those elements that are technically complex; and
  d. Adopt a work schedule to give the participants a sense of organization and create in them an urge to meet the dates set up for the completion of the different stages of the project.

Reduction of building costs is difficult to achieve unless the participants are able to contribute a significant part of the labour involved. Owing to lack of building experience, their contribution is likely to be confined to such simple tasks as excavations and transport of materials, the type of work for which labour is available fairly cheaply. This is an obvious limitation which, as mentioned earlier, can only be overcome if means are found to develop necessary skills among the participants in order to enable them to undertake most of the work required by the project.

Uncontrolled expenditure on the administration or overheads of

GASH projects can also substantially add to over-all building costs. However, because of the special nature of these projects, in which a great deal of education and training is involved, supervision expenses are likely to be higher than in other types of housing. Apart from strictly building problems, a variety of human problems within GASH programmes require sustained contribution from not only technical staff but also social workers, teachers, and administrators. This type of expenditure is unavoidable until such times when management and execution of GASH projects are adequately organized.

It is possible to cite many examples of aided self-help housing, but for every successful effort, there are many which have failed (30). An admirable effort in the field of low-cost housing is typified by the anti-malarial division of the Venezuelan Public Health Ministry which has made a remarkable contribution to building on a national scale. It is interesting to note that the division's objectives are not those of an 'agency for the construction of housing but for a training centre for the improvement of living standards in all aspects'. Technically, it concentrates on the improvement of cheap locally available materials with particular reference to stabilized soil blocks which the owner/builders can manufacture themselves.

Aid is mainly confined to the supply of materials and instructions so as to enable owners to build themselves. In some cases it has been found necessary to provide extra credit for the more skilled jobs as it was found that the quality of block laying and other specialized operations was deficient and costly when carried out by the owners. In the Venezuelan projects, a substantial part of the labour is contributed by women, children, and weekend or holiday joint community effort.

Another successful example of GASH has been reported from Chile in South America where a number of houses were built with the help of technical assistance provided by an organization known as Technicoop (the Technical Branch of the Federation of Housing Co-Operatives) (31). Finances mainly came from one of the many savings and loans associations operating in the region. A typical house has an area of 700 ft² (65 m²) and is built within the all-in cost of $2000 for workers with an average monthly income of approximately $100. Walls are made from concrete blocks and asbestos sheets are common as a wall cover.

Housing in one form or another consumes a major part of the

building resources of a country and in developing regions it presents a tremendous challenge. But in our efforts to meet the challenge of low-cost housing, it would be wrong to lose sight of the true perspective and the real nature, scope, and size of over-all building problems in developing countries of hot climatic regions. As Charles Cockburn admirably points out, 'Housing is an expression of only one kind of building demand; it omits all other types of capital investment in construction including, for example, schools, health buildings, administration buildings and roads' (32). The total building demand can only be met by improvement in the efficiency and increasing the output of local construction industry. Careful attention must be paid to the problems of contractors, extraction and manufacture of building materials, components, labour force, and professions.

At present there is a good deal of groping in this field of activity and with the best will in the world, we are still not quite certain of the lines on which it should be developed. The governments are looking for a lead in this sphere, and it is from the building professions that it must come.

REFERENCES

1. KOPPEN, W., and GEIGER, R. (1932). *Handuch der klimatologie*, vol. 4, part S, p. 76 (Berlin: Borntraeger).
2. ATKINSON, G. A. (1952). 'Building in warm climates', *BRAB Research Conference Report*, no. 5, 66–77 (Washington).
3. WEISS, E. G. A. (1963). 'Economic factors in airconditioning', *Airconditioning*, p. 102 (Brisbane: University of Queensland Press).
4. RICHARDS, S. J. (1957). 'Climate control by building design', *Proc. Symp. on Design for Tropical Living* (Durban: CSIR).
5. —— Ibid.
6. —— Ibid.
7. KOENIGSBERGER, O., and LYNN, R. (1965). *Roofs in the warm humid tropics* (London: Lund Humphries).
8. FRY, M., and DREW, J. (1964). *Tropical architecture in the dry and humid zones*, pp. 48–9 (London: Batsford).
9. GIVONI, B., and SHELDON, R. 'Influence of roof types and construction, indoor conditions in Beersheva', Research Paper no. 11.
10. SAINI, B. S. (1967). 'Durability of bush materials', *Overseas Building Notes 120* (Building Research Station, UK). Also PURUSHOTHAM, A., PANDE, J. N., and SUD, J. S. (1963). 'A note on fire-resistance-cum-antiseptic composition and fire-resistant paint', *Journal of the Timber Development and Product Association, India*, **9**, 3.
11. NARAYANAMURTI, D. (1959). 'Building boards from indigenous materials', *Proc. Symp. on Timber and Allied Products* (New Delhi: National Buildings Organisation).

12. MIDDLETON, G. F. (1952). 'Earth wall construction', *Commonwealth Experimental Building Station Bulletin*, 5 (Sydney).
13. UNITED NATIONS (1964). *Soil-cement, its use in building* (New York).
14. HUDSON, J. C. (1947). 'The corrosion of iron and steel and its prevention'. *Jl Oil Colour Chem. Assoc.* **30**, 35 cited by HOLMES, B. M. (1951). 'Weathering in the tropics', *CSIRO Report* 41 (Melbourne, restricted circulation).
15. HOLMES, B. M. (1952). 'The weathering and durability of building materials under tropical conditions', *Building Research Congress Proc.* (London).
16. WHITLEY, P. (1958). 'Paint problems in tropical climate', *Overseas Building Notes*, no. 55 (Building Research Station, UK).
17. COMMONWEALTH EXPERIMENTAL BUILDING STATION. 'Concrete bricks and blocks', *Notes on science of building*, **14** (Sydney).
18. FOOTNER, H. B., and MURRAY, G. (1947). 'Tropical paints', *J. Oil Colour Chem. Assoc.* **30**, 378–90.
19. BROOKS, C. E. P. (1946). 'Climate and the deterioration of materials', *Quart. J. R. Met. Soc.* **72**, 87–97.
20. CSIRO DIVISION OF FOREST PRODUCTS. 'Eradication of termites', Reprint Newsletter (Melbourne). See also CEBS, 'Termite proofing of buildings in Australia', *Notes on science of building*, **11** (Sydney).
21. ROBERTS, P. W. (1962). 'Desert areas: dry hot conditions', *Civil Engineering*, **32** (2), 46–9.
22. MACPHERSON, R. K. (1961). 'Environmental problems in tropical Australia', Commonwealth of Australia Report (reprint). Also LAMBERT, G. E. (1961). 'Work, sleep, comfort', *Environmental physiology and psychology in arid environment* (Paris: UNESCO).
23. LAMBERT, G. E. Ibid.
24. COCKBURN, C. (1963). 'Building economics—India', *Interbuild*, **10** (8), 9–15. See also MUSTAFA, A. M. (1968). *Investigation of the constructional techniques of building in the arid regions of Australia*, unpublished MArch thesis, University of Melbourne.
25. TURIN, D. (1969). 'The economic significance of construction', *Architects' J.* **150** (42), 923–9.
26. SAINI, B. S. (1968). 'Industrialised building in a remote area', *Arena-Interbuild*, **83**, 39–43. See also SAINI, B. S. (1969). 'Industrialised building in tropical Australia', *Architecture in Australia*, **58**, 827–33 (Sydney).
27. JENNINGS, V. E. (1966–7). 'Building in locations remote from sufficiently large urban communities', lectures to Post-graduate Course in Tropical Architecture, University of Melbourne.
28. ALCOCK, A. E. S. (1966). 'Self-help housing in South East Asia', *Australian Planning Institute Journal*, **4**, 146–52.
29. UNITED NATIONS (1964). *Manual on self-help housing* (New York). A good general introduction to self-help housing can be found in EYHERALDE, R. F. (1957). *Self-help housing* (New Delhi: Ford Foundation).
30. TURNER, J. C. (1963). 'Dwelling resources in South America', *Architectural Design*, **33**, 360–93.
31. —— Ibid.
32. COCKBURN, C. (1969). 'Construction: a management problem', *Architects' J.* **150** (42), 930–1.

8

Principles of structural fire-resistance

H. L. Malhotra, B Sc (Eng), C Eng, M I C E,
M I Fire E
Head of the Structural and Materials Fire Test Section,
Fire Research Station, Boreham Wood

8.1. Introduction

8.2. Brief history of statutory control

8.3. Pattern of a fire

8.4. Aim of structural fire protection

8.5. Fire-resistance and its determination
8.5.1. General
8.5.2. Stability
8.5.3. Specimen size
8.5.4. Apparatus

8.6. Properties of materials at high temperatures
8.6.1. General
8.6.2. Steel
8.6.3. Aluminium
8.6.4. Concrete
8.6.5. Wood
8.6.6. Plastics

8.7. Behaviour of structures in a fire

8.8. Design features of structural elements
8.8.1. Beams
8.8.2. Floors
8.8.3. Walls
8.8.4. Columns

8.9. Future trends in fire protection

8.1. INTRODUCTION

Serious fires have occurred in the past but the continuously increasing and serious economic losses they cause can be associated with the industrial growth of a country, particularly where this has resulted in a high density of industrial and commercial premises. The increasing sophistication of machinery and the tendency to undertake processes in vast open areas in buildings has the effect of increasing the potential loss from a single fire to millions of pounds. It is hoped that fire losses are not accepted as a price to be paid for the industrial development of a country.

In the UK records have been kept since 1947 (Fig. 8.1) of the annual fire losses and it is apparent that there has been a steady increase both in the number of incidents and the direct losses. In 1968 (1), the last year for which records are available there were over 200000 fires attended by the fire brigades and the direct losses exceeded £100 million. Whilst no reliable information is available on consequential losses, rough estimates show that these may add another £200 million to the bill. The country's investment in fire protection amounts to approximately £200 million; this includes the cost of fire protection measures in buildings, fire research, and the running of the fire brigades.

The need for measures to protect life and property has long been recognized but until the last decade or so many of the requirements, though sensibly based, had no scientific foundation. Fire technology is a new discipline and some important advances have been made in this field since the early 1950s when fire research became a recognized scientific activity. The findings of research are used to rationalize and improve the requirements of many authorities who have the responsibility for fire safety in buildings.

Measures that are specified for fire protection in buildings include design, means for the safe evacuation of people, fire detection and warning devices, and first-aid devices such as sprinklers. Until recently these have been regarded as separate aspects of fire protection considered in isolation from each other. For a realistic and logical

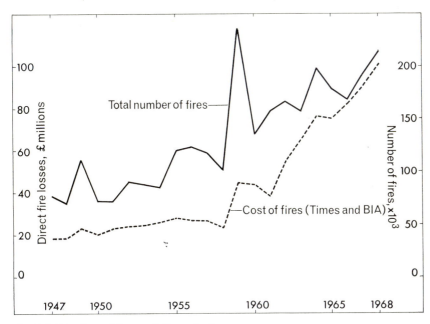

Fig. 8.1. Number of fires and direct fire losses

assessment of the value of fire protection measures due cognizance must be taken of their interrelation. This chapter deals only with the design principles for structural elements that are provided to control fires on the concept of containment and the recent developments in this field.

8.2. BRIEF HISTORY OF STATUTORY CONTROL

Ever since the Great Fire of London there has been concern over the design of buildings to resist the effects of fire; the City of London promulgated some ordinances to deal with this problem. It is only over the last hundred years or so that general control has been exercised by the local corporations or authorities under Acts of Parliament. A Public Health Act in 1848 gave discretionary powers of limited scope to the local corporations for the safe and sanitary development of towns and after ten years it was amended requiring the local authorities to make bylaws for this purpose. A section of the Act required the walls of the buildings to be structurally stable and capable of withstanding fire, indicating concern with the spread of fire from one building to another. In 1875 this section was re-enacted widening the scope of the requirements and remained the principal basis for bylaws until 1936, when as a result of a departmental committee report, a new Public Health Act was passed. This required the local authorities to make bylaws on the specific subject of public health and safety and in 1961 these powers were transferred to the Minister of Housing and Local Government.

A technical committee of the Ministry of Works examined the problems of fire protection in buildings and produced in 1946 a report on fire grading of buildings. Part 1 (2) dealt with the general principles and the structural precautions and formed the foundation for the Model Bylaws issued by the Ministry of Housing and Local Government in 1952/3. These were used by the local authorities in drafting their own bylaws. By virtue of the amending legislation in 1960–1, building regulations were issued in 1963 for Scotland (3), followed by those for England and Wales (4) in 1965. The two bodies responsible for the regulations (the Scottish Development Department and the Building Regulations Advisory Committee) have under continuous scrutiny new knowledge on fire protection and periodic amendments are made to the regulations.

London occupies a unique position in having its own constructional bylaws under an Act of Parliament. These were first made in 1939 and have been since replaced by the 1964 bylaws which are applicable to the Central London area.

8.3. PATTERN OF A FIRE

The course of an unattended fire consists of three distinct stages, a growth period, the steady combustion period, and the decay period. The majority of fires develop from a small source, such as the ignition of a piece of furnishing in a house, a waste-basket in an office, a container in a warehouse, and waste material in a factory. If the conditions are favourable, i.e. adequate supply of air and combustible material is available, the fire will spread to adjacent items accompanied by a progressive build-up of heat. This causes the burning materials to decompose at an accelerated rate leading to further rise in the temperature of the environment and the spread of fire. If the wall and the ceiling surfaces become involved the fire spread can occur at a faster rate contributed by the elongation of flames at the ceiling level. The duration of this growth stage depends upon the nature of materials involved and other environmental factors.

When the general level of temperature in the compartment exceeds about 250 °C, the combustible materials begin to generate flammable vapours, the continued rise of temperature to above 600 °C leads to the rapid involvement of all contents the phenomenon being referred to as a 'flashover'. After this the steady combustion phase follows with temperatures maintained at a high level, around 1000 °C, and it is during this phase that the structural fabric of the building is subjected to severe heating conditions and called upon to play its part in containing the fire. The duration of this stage depends upon the total amount of the combustible materials and the rate at which they can decompose. If no effort is made to control the fire in this stage then, after most of the materials have been consumed, the fire reduces in intensity with the progressive lowering of temperatures to the ambient. A well-designed building can withstand the complete burn-out of the contents without suffering collapse or permitting the fire to escape from the protected areas.

Some recent studies (5) have shown the influence of various factors on the rate of development of fire and its severity. Some of the important ones are the fire load density (i.e. the amount and the distribution of combustible contents), openings through which air can be supplied, the geometry of the building, the nature of the exposed surfaces, and the thermal properties of the compartment boundaries. As a result of these studies it is possible to predict the expected severity of fires in buildings more realistically.

8.4. AIM OF STRUCTURAL FIRE PROTECTION

The aim of measures for structural fire protection is to ensure that for a given severity of fire, the elements which constitute the building are so designed as to continue to fulfil their assigned functions. Structural precautions form the main basis of statutory control exercised for fire safety in buildings. The bases for the requirements need to be understood by the architects and designers so that the provisions may be effective as well as economically justifiable. When no consideration has been given to the fire protection aspects during the early stages, modifications to improve the design at a later stage can be an expensive and time-consuming business. If the fire protection requirements are included amongst the other design criteria at an early stage, they can be provided easily and often advantage can be taken of the associated improvements in thermal and sound insulation.

Data on the methods of achieving structural fire protection are given in the 'deemed to satisfy' provisions in the Building Regulations (3, 4) and Bylaws (6) as well as some of the structural Codes of Practice (7). Much of the general information provided is supplemented in the catalogues of manufacturers dealing with products used in this connection. Publications (8, 9) are also made by the Fire Research Station to provide authoritative data on various forms of constructions and some of the current research work should lead to the publication of design guides.

8.5. FIRE-RESISTANCE AND ITS DETERMINATION

8.5.1. General

During the early part of the century Ingberg (10) in the USA carried out experiments on fires in buildings, on the basis of which he proposed a time/temperature relationship for determining the fire resistance. Quite independently work was carried out by the British Fire Prevention Committee at their laboratories in Regent's Park which led to similar suggestions being put forward to the British Standards Association (now the British Standards Institution) for the formulation of a standard testing procedure. This was issued in 1932 as BS 476 (11): *Tests for the incombustibility and fire resistance of building materials and structures.* The time/temperature curve specified in this standard is shown in Fig. 8.2 and this remained unchanged when a revised version was produced in 1953 as BS 476: Part 1: *Fire tests on building materials and structures* (12). Since then research has been carried out on the conditions that exist in building fires which show that higher temperatures can occur during the early stages of a fully developed fire and the flames are characterized by high emmissivity conducive to rapid heat transfer by radiation. The peak temperatures in fires are not sustained for long and the over-all severity of exposure can be related to the furnace tests. It has not been considered necessary or practicable to alter the shape of the standard curve, as it is used with only minor differences by many countries, and much knowledge has been acquired on the basis of this relationship. It has not been shown to have created lack of safety in buildings.

In 1961 a technical committee of the International Standards Organisation (ISO/TC92) was set up to produce an internationally acceptable standard for conducting fire-resistance tests. All the leading international laboratories participated in the work of this committee and an agreed procedure was published in 1968 as the International Standards Recommendation No. 834 (13). In Fig. 8.2 the time/temperature relationship of this recommendation is compared with the curve specified in BS 476: Part 1: 1953. The differences are marginal, the international curve possessing the advantage of being expressed as a mathematical relationship, as follows:

$$T - T_0 = 345 \log_{10} (8t + 1)$$

where T = temperature in °C after time t,

T_0 = initial ambient temperature in °C,

t = time in minutes.

The British Standard has recently been revised[1] and full use has been made of the agreements reached in the international field. The new standard is being published as BS 476: Part 8: *Fire resistance tests for elements of building construction* (14). The description of the testing procedures and the concepts which follow are based on the specifications contained in the revised standard.

Fire-resistance is defined as the ability of an element of building construction to fulfil its assigned function in the event of a fire. It is not the property of an individual material but the behaviour of a recognizable constructional element which may consist of a single material or may be fabricated from a number of different materials.

Fire-resistance is measured as the time for which a prototype

1. At the time of writing the publication date was expected to be early 1971.

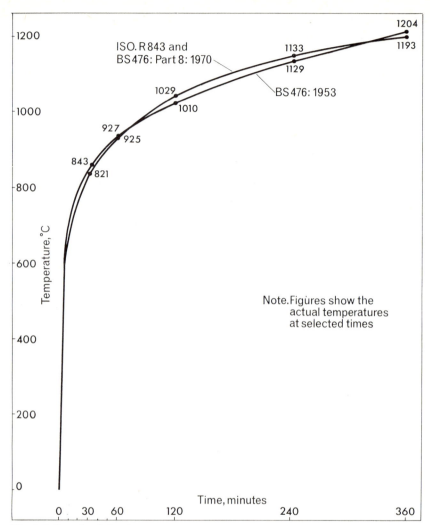

Fig. 8.2. Standard time/temperature curves used for fire-resistance tests

sample of the construction will withstand a specified heating programme and satisfy the appropriate criteria. The heating programme
simulates a standard fire and the test criteria take into account the
functions of the building element.

The standard specifies criteria for stability, integrity, and insulation.
a. 'Stability' implies freedom from collapse.
b. 'Integrity' implies freedom from the development of orifices or openings which can allow the passage of hot gases and flames.
c. 'Insulation' requires the structure to posses sufficient resistance to the passage of heat when separating spaces so as to prevent the ignition of combustible materials in contact with the unheated face.

For the main types of elements in a building, Table 8.1 shows the application of various criteria.

Table 8.1. *Application of fire-resistance criteria*

Building element	Stability	Integrity	Insulation
Column	×	—	—
Beam	×	—	—
Wall	×	×	×
Floor	×	×	×

Note. × means applicable; — means not applicable.

The latest standard has moved away from the concept of rigid classification or grading on the basis of performance, instead the times for which a construction complies with the various criteria are shown. This should encourage flexibility in the use of data as in some cases relaxations could be made for some of the criteria provided the situation justifies this. The Building Regulations will for the time being continue to specify their requirements in terms of the existing grading periods, the future amendments could well make use of the flexibility by choosing some intermediate values where these seem to be more appropriate.

8.5.2. Stability

Freedom from collapse under conditions of fire is taken to imply that not only will a building remain as an integral unit during the course of a fire but even afterwards it is not likely to collapse thereby hindering the salvage operations. The continued stability of the construction was ensured in the past by maintaining the test load for a period of forty-eight hours after the end of heating. This arbitrary period was chosen to take account of any reduction in strength during the cooling down stage.

The 1932 version of the BS specified that during a test the construction should be subjected to $1\frac{1}{2}$ times the maximum permissible design load. In practice the loads in buildings only rarely reach the maximum design values hence the requirement to subject samples to $1\frac{1}{2}$ times the design load was considered unrealistic and modified in 1953 to the design load. However, the maintenance of the load for forty-eight hours or its re-application necessitated termination of the heating period at a selected time when it could be safely assessed that the residual strength after cooling would be adequate. In practice this creates a tendency to 'play safe' without establishing the full potential of a construction.

In the latest revision of the standard a new approach is being tried out which includes the residual strength consideration of the previous standards but permits the establishment of the full potential of the specimen construction. Figure 8.3 shows a hypothetical case of the reduction in strength of a specimen construction when subjected to a fire test. The ultimate strength is gradually lowered from M_U until at time t_1 it is reduced to the same value as the applied moment M_A and collapse occurs. If heating is terminated at an earlier time t_2, any further reduction of strength occurs at a slower rate but if the load is maintained structural failure will take place at time t_3, i.e. (t_3-t_2) after the end of heating, when the strength curve crosses the M_A ordinate. If, however, the heating terminates at time t_4, the final residual strength will not be reduced below the value of M_A, hence the construction will retain its stability and continue to support the applied loads. The aim of testing should be to establish time t_4 as closely as possible. The new standard achieves this by proposing that tests may be carried out to the point of collapse (time t_1) and the time t_4 established using the arbitrary relationship $t_4 = 0.8\,t_1$.

It is seen from Fig. 8.3 that the duration for which the construction can successfully withstand collapse under fire conditions is partly dependent upon the value of M_A, the applied moments, or the stresses generated in the construction by the applied loads. Normally, constructions are tested under full load conditions but there may be circumstances where lower loads are justified owing to the specific use to which they will be put in the building provided there is no likelihood of a later change of use resulting in any increase in live loads.

During the course of a fire, the actual loads to which elements are

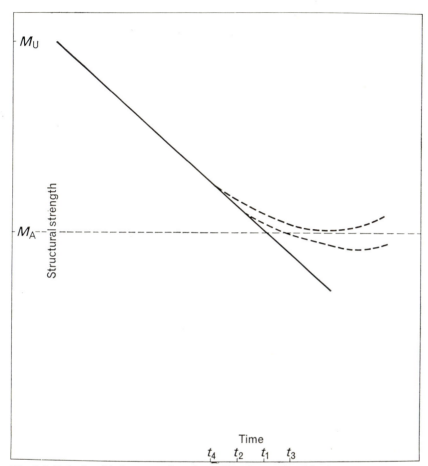

Fig. 8.3. Relationship between duration of heating and residual strength

subjected and the magnitude and the nature of stresses developed in them can vary. Restrictions to the free movement of elements to accommodate thermal expansion by the built-in structural restraint can induce high local stresses, continuity at junctions can lead to reversal of stresses and partial failure of one or other element can lead to a redistribution of stresses. To predict the nature and the magnitude of these changes is not possible with present knowledge on the subject, hence the laboratory procedure makes assumptions on the constancy of imposed loads and the nature of stresses. Studies are in progress on this subject and the test procedures may change in future to take account of the load variability. Currently the standard requires the boundary conditions for the specimens to represent the actual use but where these cannot be accurately specified, as in the case of the majority of constructions, certain conditions are arbitrarily assumed and reproduced in the tests, as shown in Table 8.2.

Table 8.2. *Boundary conditions for test elements*

Type of element	End or edge conditions
Column	Restrained in direction at both ends
Beam	Simply supported at the ends
Wall, load-bearing	Restrained in direction at top and base, free along vertical edges
Wall, non-load-bearing	Restrained along four edges
Floor	Simply supported on two opposite edges, other two free

The imposed loads are assumed to remain the same, therefore columns and walls are allowed to undergo a vertical movement in order to maintain a constant load.

Structural failure occurs when the applied loads cannot be maintained, in addition the new proposals require that for beams and floors, the loss of stability is imminent when the vertical downward deflection exceeds $l/30$ where l is the clear span.

8.5.3. Specimen size

The size of the specimen used in a test has an important bearing on the application of data to buildings. The specimen must be of a large enough size to reproduce the realistic use of materials and the nature and the magnitude of the different types of stresses which live loads will develop in the actual construction. The modelling of fire-resistance tests to a small scale is not yet possible as the appropriate scaling laws have not been established. There is a limit to the reduction in size below which the materials will not behave as in a larger construction, for example some years ago in the course of an investigation (15) on prestressed concrete beams, models scaled down to quarter size were

Table 8.3. *Specimen sizes*

Type of element	BS 476 (minimum requirements)	Fire Research Station (maximum possible)
Column	3 m	3 m
Beam	2·5 m	3 m or 7·5 m*
Wall	2·5 × 2·5 m	3 × 3 m
Floor	2·5 × 4 m	3·5 × 4 m or 3·5 × 7·5 m*

*The equipment can be adapted for these larger sizes where necessary.

prepared. The results given by the reduced scale specimens did not fully agree with the full-size constructions owing to differences in the failure pattern.

Furnaces for conducting fire-resistance tests are expensive and require large laboratory space. It is therefore not possible to have the equipment of a size capable of dealing with elements intended for use in industrial and commercial buildings; the size usually selected for test purposes is that appropriate to dwellings. It has been found that the test data can be confidently used in the design of larger elements for other types of buildings. Table 8.3 shows the minimum sizes given in the BS and the capacity of the equipment installed at the laboratories of the Fire Research Station at Boreham Wood.

8.5.4. Apparatus

The apparatus used at the Fire Research Station for the tests comprises three gas-fired furnaces; a horizontal one for floors and beams, a vertical one for walls and partitions, and a cylindrical furnace in two halves for testing columns. The furnaces use a mixture of town gas and air as fuel and have a large number of burners to give a uniformly heated environment. The heat transfer to the specimen constructions is a combination of convection and radiation; owing to the non-luminous nature of flames, convection is predominant during the early part of the test and as the furnace walls become hot, radiative transfer becomes dominant.

One of the main differences between the heating conditions in an actual fire and the furnace tests is in the nature of the environment. The luminous flames in fires are more radiative so that in a fire there

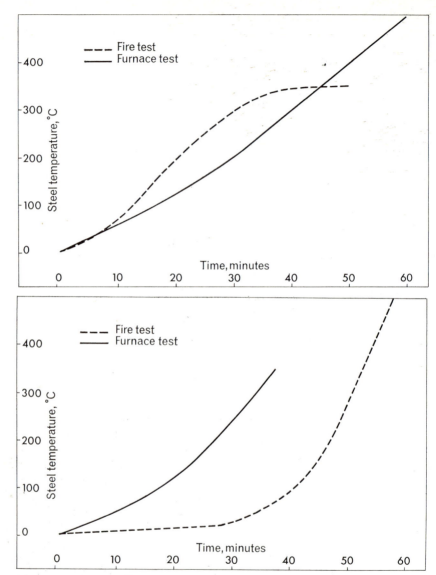

Fig. 8.4. Comparison of steel temperatures in fires and in furnace tests
(*a*) Rapid growth fire (16)
(*b*) Slow growth fire (17)

Fig. 8.5. Furnace equipment for testing load-bearing walls. Photograph shows the specimen in the loading frame and the lighting of the furnace

Fig. 8.6. A prestressed concrete beam after collapse. The beam is located on top of the horizontal furnace and is loaded hydraulically

will be a high heat transfer coefficient even during the early stages. After some time the differences between the characteristics of the flames in fires and the environment in the furnaces become small.

Some comparisons have been made between the temperatures attained by protected steel sections in fires and in the furnace tests during the course of two recent investigations. When the development of the actual fire was rapid, the temperature attained by the steel sections (16) was considerably higher than in the furnace until the fire passed its peak; in the example shown in Fig. 8.4 the same temperatures were reached after 45 minutes with a fire load density[1] of 30 kg/m². In the other investigation (17) concerned with fires in dwellings, temperatures rose more slowly and hence the temperature rise of the steel sections was at a very low rate until the fire reached the flashover conditions. After this the steel temperature rose rapidly in the fire, at nearly twice the rate in the furnace test (Fig. 8.4).

These examples show that the conditions in actual fires can differ widely and, during the time when high temperatures and high emmissivity flames exist, heat transfer to a construction is at a higher rate than in the furnace tests. However, the furnace test procedure has better reproducibility, more control can be exercised on the heating programme and it provides sustained high temperature conditions in comparison with actual fires.

The effect of furnace characteristics on the performance of structural elements is being studied by a commission of the Conseil International du Batiment on fire research. The first aim is to establish the degree of correlation between the furnaces in different countries and secondly to draw up a specification on furnace characteristics to ensure reproducible conditions in different countries.

1. Fire load density is expressed as the amount of timber per unit floor area having the same calorific value as the combustible contents.

8.6. PROPERTIES OF MATERIALS AT HIGH TEMPERATURES

8.6.1. General

A knowledge of the properties of structural materials at high temperatures is essential for an assessment of the performance of structures under fire conditions and to make an optimum use of the different materials available for their construction. Materials undergo physical and chemical changes as they are heated, with organic products the chemical changes can result in rapid decomposition and active combustion. Steel, aluminium, concrete, timber, and plastics are used structurally in buildings; of these, plastics have only recently entered this field and their use is at present limited to a few special situations. The most important property on which data are required is that of strength and its variation with temperature.

Fire temperatures have been known to approach and to occasionally exceed 1200 °C, the standard curve of the BS has a value of $\simeq$ 1200 °C at six hours, though in practice under the building regulations for the majority of buildings the fire requirements do not exceed two hours and therefore only temperatures up to 1000 °C are generally of interest. In practice, however, it will be found that strength at temperatures in excess of 800 °C is so low that the majority of building materials have only a nominal structural contribution to make. The integrity of materials at elevated temperatures has also an important bearing on fire-resistance, the ability of the outer layers to stay in place assists in keeping the temperature of the interior of a structure low, thereby enabling it to give a good performance.

Materials expand as their temperature is raised and if any restraint is provided to free expansion, internal stresses are developed leading to distortion. Provision for expansion is an important consideration for metals. There are many instances of unprotected steel beams pushing out the supporting walls or when unable to do so, suffering sever deformation.

The other changes which take place are an increase in thermal conduction, a decrease in heat capacity, physical disruption of certain materials, and the decomposition and burning of others. Some of the important physical properties are listed in Table 8.4 and the others are discussed under individual materials.

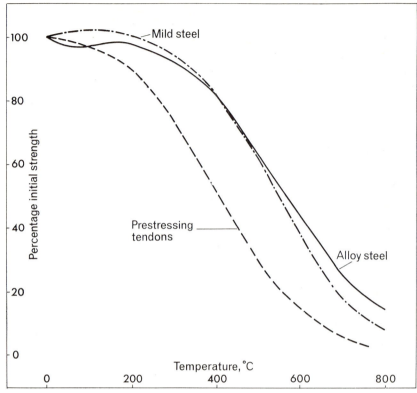

Fig. 8.7. Strength of steel at high temperatures

Table 8.4. *Thermal properties of materials*

Material	Density kg/m³	Coefficient of expansion/°C	Specific heat J/kg°C	Thermal conductivity W/m°C
Steel	7800	11×10^{-6}	440	42
Aluminium	2280	29×10^{-6}	870	230
Concrete	2400	13×10^{-6}	1120	1·5
Timber	530	—	1740	0·15
Plastics	900–2500	70×10^{-6}	1800	0·12–4·0

8.6.2. Steel

Fig. 8.7 shows the strength of steel of various types at temperatures up to 800 °C. Mild steel shows a slight gain in strength initially, followed by a fairly rapid loss until 700 °C. By 550 °C its strength is reduced by approximately 50 per cent and on the assumption that a factor of safety of 2 exists in structural design this has been generally quoted as its 'critical' temperature. The high-strength alloy steels follow a similar pattern when heated. Cold-worked high-strength steels on the other hand show a markedly greater reduction in strength and the 'critical' temperature for steels used for prestressing tendons has been shown to be in the 400–450 °C range.

Mild steel and alloy steels after cooling regain almost all of their initial strength but cold-worked steels do not. They suffer a permanent reduction in strength probably because of the reversion to the original mild steel structure. Structures using such steels are therefore likely to show permanent deformation even after fires of short duration, if their temperature has exceeded 200 °C.

8.6.3. Aluminium

Aluminium is not often used structurally in the design of buildings probably owing to its higher cost and lower strength in comparison with steel. It loses strength at a more rapid rate, a reduction of 50 per cent in initial strength can take place at temperatures just in excess of 300 °C. As the temperature is raised it begins to soften, the melting points depends upon the type of alloy used but is generally in the region of 650 °C. This means that unprotected and exposed sections of aluminium will melt even during the course of a thirty-minute fire test when the temperatures reach nearly 850 °C.

Aluminium sections are generally employed in the fabrication of lightweight partitions, they need to be protected and provided with facilities to expand in view of the high coefficient of thermal expansion of the material. Outside the building field the material finds an important use in the construction of load-bearing superstructure of ships and it is invariably protected against the effects of fire.

8.6.4. Concrete

Concrete is a heterogenous material consisting of two main constituents, cement paste and aggregate. It is an important structural material which by itself or in combination with steel is to be found in a large number of buildings. A number of investigations (18–20) have been made since 1920 on the effect of high temperatures on its compressive strength. Much of the work has been with concrete made with gravel aggregates though some of the more recent investigations have examined limestone aggregate and the lightweight aggregate materials. On raising the temperature of concrete a reduction in its compressive strength begins—soon after 100 °C and by 450 °C (Fig. 8.8), the gravel aggregate concrete may lose as much as 50 per cent of its initial strength. If concrete is maintained at this temperature or allowed to cool down, its final residual strength can be 15–20 per cent lower than that at the elevated temperatures. Unlike steel, the strength loss phenomenon for concrete is non-reversible, probably to a large extent due to the chemical changes in the hardened cement paste. The cracking or splitting of aggregate is also likely to contribute to the permanent lowering of the strength. Recent experiments with other aggregates shows that lightweight materials lose strength at a lower rate than the natural rock or stones.

In addition to loss in strength at high temperatures concretes of certain types are also susceptible to damage by spalling. The exact nature of this phenomenon is not yet fully understood but it is caused when a structure is exposed suddenly to high temperatures causing the development of steep thermal gradients across the section. The inducement of high thermal stresses below the surface, the unequal expansion between the aggregate and the matrix of cement paste and the evaporation of moisture causing a vapour pressure in the pores all contribute to spalling. In its severe form pieces of concrete fly off from the surface with an explosive force, generally during the initial stages of heating, or bits of concrete become dislodged from corners and faces during the later stages. The concrete most susceptible to spalling is that of a dense nature made with silicious aggregates. As gravel contains over 90 per cent silica, it is more prone to this effect, whereas concretes made with limestone and lightweight aggregate usually do not spall.

As an alternative to using special aggregates, the effects of spalling can be minimized by the introduction of supplementary reinforcement below the exposed surfaces. Such reinforcement may consist of a wire fabric or a network of closely spaced links as near to the surface as

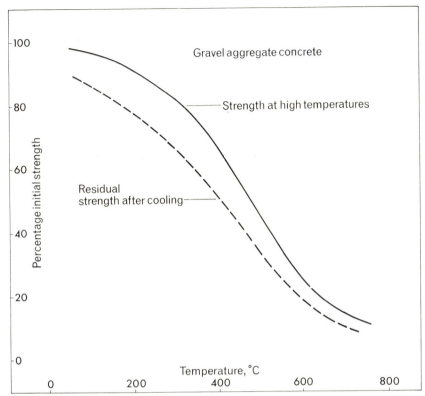

Fig. 8.8. Strength of concrete at high temperatures

practicable. The reinforcement cannot prevent the tendency of the concrete cover to break up but retains it in place and if spalling does take place, it cannot proceed beyond the reinforcement. A more effective way is to provide an insulating coating on the surface which, by reducing the surface temperature and consequently the thermal gradient across the concrete section, prevents conditions favourable for spalling.

The role of concrete in structures is not only as a load-bearing material but under fire conditions it can also provide thermal insulation. Concrete cover in a reinforced or prestressed structure has this important function to serve and therefore concretes of low density have an advantage over the denser materials. The thermal conductivity of lightweight concrete may be as much as 25 per cent lower than the denser material.

8.6.5. Wood

In fairly thin sections wood is easily ignited and burns readily in a fire, particularly if all the surfaces are exposed. However, with sections of 25 mm or more in thickness, once the surface has become charred, any further damage occurs at a slow and steady rate. This phenomenon is due to the formation of charcoal on the surface and the low conductivity of the material. Timber having a density of 400–650 kg/m³ has been shown to char at an average rate of 0·5–0·6 mm/min when used as a beam but at a slightly higher rate when used as a column.

The low conductivity of wood results in a clear demarcation between the charred and the uncharred section. There is some reduction in the ultimate strength of wood as its temperature is raised above 100 °C, in some tests on laminated columns this was shown to be under 20 per cent. The predictability of charring of timber is a useful guide in assessing the fire performance of structural timber elements and data are available for floors and beams (21, 22). For very slender members some further data are being obtained to establish the effects of increasing slenderness ratios.

Lamination of timber has meant that timber sections with predictable strength properties can be designed and large sections can be fabricated with controlled amounts of natural flaws. Investigations (23) have shown that the choice of glue is important as certain types, such as casein, can lead to the separation of laminates before they have been charred to the glue line. Phenolic and resorcinol glues are the best products to use.

Recently interest has been shown in the use of timber sections to clad steel stanchions and beams, both for appearance and to protect them from the effects of fire. The design of such a system requires careful attention as shrinkage at joints and corners could lead to early failure.

0   20   40   60   80 millimetres

0   2   4   6   8   10   12 inches

Top   Centre   Bottom

Fig. 8.9. Laminated timber column before and after fire exposure. Cross-sections illustrate the typical pattern of charring obtained with timber

8.6.6. Plastics

As a building material plastics have not made great inroads into the structural field. The viability of an all-plastics house is in some doubt owing to concern about its fire behaviour and lack of data on durability. The organic nature of plastics means that decomposition occurs at temperatures low in comparison with those experienced in fires, and the high conductivity, except for foamed materials, results in rapid heat transmission. Hence in all exposed situations plastics materials are likely to suffer serious damage under fire conditions. The behaviour of plastics in a fire varies from the highly flammable styrenes to the low flammability phenols and vinyls. It is now possible to make most of the plastics resistant to ignition but their decomposition at elevated temperatures cannot so far be successfully controlled.

To be structurally employed plastics products need to be protected against the effects of heat; this may have the effect of restricting their use on economic grounds. Reinforced resins have been proposed for the construction of large panels, in a multistorey building in London (24), the external walls are being prefabricated complete with internal finishes and hoisted into place. These are attached to the steel framework and they serve the function of a complete cladding. Foamed plastics can be bonded to skins of steel and used in the cladding of industrial buildings. These composite panels possess high strength under normal conditions, the effect of fire will be to weaken the bond between the steel facing and the core. By using a fixing method which does not rely on the bond of the sandwich panel, such constructions can be successfully employed.

One of the ways in which foamed plastics may be used is in combination with traditional materials to provide low density products. Expanded polystyrene aggregate with cement is a useful way of producing lightweight blocks and some investigations are in progress to establish the fire performance of such combinations.

8.7. BEHAVIOUR OF STRUCTURES IN FIRE

The design criteria in the structural field are being based more and more on establishing the failure pattern under a given set of conditions, with the result that a 'limit state' concept is applied. Limit state implies that the structure will not become unfit for the use for which it is required within the acceptable margins of safety. It has been suggested that fire may be considered as an accidental overload; whilst this may be true in a few cases as a generalization it does not embrace the complex ways in which a fire can affect the behaviour of a structure, nor is it possible to take into account phenomena such as spalling causing premature collapse.

In some instances the limit state concept could be applied using the parameters for the properties of materials at ambient and high temperatures established by tests on prototype constructions. As an example, a simply supported beam with uniform loading will have certain applied moments developed due to the live load as shown in Fig. 8.10, with a maximum value of M_A. The design of the beams may be such that with the materials used it has an ultimate moment capacity of M_U. The ratio between M_A and M_U represents the factor of safety. As the beam is heated the materials suffer a reduction in their strength properties and this leads to a lowering of the ultimate moment resistance M_U and a reduction in the factor of safety. As long as M_U has a value greater than M_A the beam will continue to support the applied load. When the two become the same the limit state for stability has been reached and structural collapse will ensue.

One of the aims of the design of structural elements which are likely to be involved in fires is to prevent the limit state for stability being reached under the severity of fire to be expected, or in other words to ensure that the factor of safety will always remain in excess of unity.

Structural elements to be found in buildings can be divided into four main types, viz. beams, columns, floors, and walls. Of these beams and columns have a loadbearing function only whereas the other two also serve as a fire separation. There is a similarity in the collapse pattern for beams and floors as well as between columns and walls. In terms of heat transfer the elements represent different situations, with walls and floors heat transfer takes place from one face only, beams usually have three faces exposed and free-standing columns four. Beams and columns may in some situations be so erected that parts are shielded from direct exposure.

Under fire conditions the exposed surfaces of the structural elements attain high temperatures and transfer of heat takes place to the

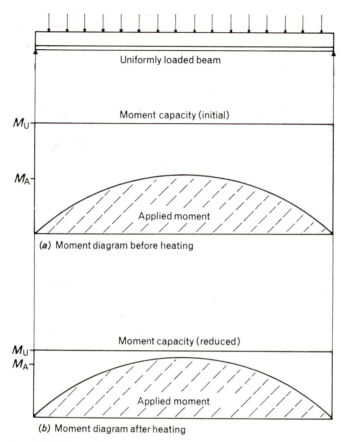

Fig. 8.10. Effect of fire on moment capacity of a simply supported beam
(*a*) Moment diagram before heating (*b*) Moment diagram after heating

interior of the section. With beams and floors the lower parts become
heated at a quicker rate than the rest of the construction and these
parts form the tensile zone of the structure, hence it is the behaviour
of this zone which determines the stability of the construction. With
homogenous material, such as steel and timber the resistance capacity
of the tensile zone is gradually decreased by loss in strength of steel or
by the erosion of timber from the surfaces. With heterogeneous concrete
structures the ultimate behaviour is determined by the performance

of the steel bars or tendons. This plain analysis is appropriate only for simply supported structures or others which behave in a similar manner, but where the boundary conditions are such that significant restraint against movement is provided, the thermal stresses lead to a redistribution of stress pattern within the structure. Depending upon the nature of the structure and the magnitude of the thermal stresses, a construction may show a marked improvement in its performance owing to repositioning of the neutral axis or the secondary effects of thermal stresses may cause earlier failure by local disruption of materials. Present knowledge does not permit a true analysis to be made of composite structures and researches are being carried out to obtain data.

The majority of walls and columns are under direct compression and the effect of heat is to lower the resistance capacity by the gradual weakening of materials from the heated faces. This will eventually lead to a compression failure of the element when the working load achieves the significance of the ultimate load for the weakened section. However, in practice, often the increase in the slenderness ratio caused by the progressive weakening of the sections results in the development of bending stresses, accentuated for walls by the deformation produced by the unequal expansion of the heated and the unheated faces, thus making the analysis of the behaviour of walls and slender columns more complex.

To analyse the behaviour of a structural element under fire conditions it is necessary to have a knowledge of three aspects:

a. Temperature distribution within the section.

b. Properties of materials at high temperatures.

c. The failure pattern for the element.

The standard fire as represented by the time/temperature relationship in Fig. 8.2, is an unsteady-state heating and the factors which determine the heat transfer into the section are the thermal conductivity, the density and the specific heat of the materials. For a one-dimensional heat flow into a homogeneous material, the rate of heat flow dQ/dt is a function of $K/\rho C$,

where ρ = density of material,

C = specific heat of material,

K = thermal conductivity.

Factor $K/\rho C$ is called the thermal diffusivity of the material.

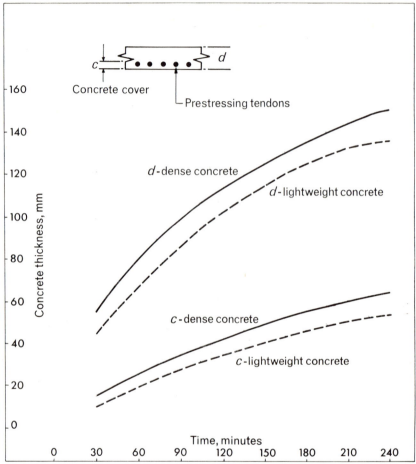

Fig. 8.11. Fire-resistance of prestressed concrete slab floors

Solutions have been provided for simple cases, when structures are subjected to the heating conditions represented by Fig. 8.2, by assuming equivalent constant temperatures for different periods of heating (25). Two- and three-dimensional heat flows are more difficult to analyse particularly for slender sections but some approximate solutions are possible.

As an example, it is possible to calculate the thickness of concrete cover in flat decks of prestressed concrete where the time for the temperature of the tendons to reach 400 °C is taken as the limit for stability. This permits curves to be drawn as shown in Fig. 8.11 for two types of concrete; the thermal diffusivity values for concretes were derived from the data from actual fire tests. Information of this type has been used in the preparation of tables for the section on fire protection in the new Code of Practice on the structural use of concrete in buildings (7).

Some data are available on the strength of concrete and steel at high temperature, but no systematic studies appear to have been made to determined the variation in conductivity, specific heat and density of these materials. To enable the wider use of analytical methods the properties of the normal range of materials need to be established at various temperatures up to 1000 °C.

The knowledge of the failure pattern for different types of constructions can be obtained only from the observations made in fires and in full-scale fire-resistance tests. For some simple forms of constructions such data are already available and for others there is need to undertake investigations designed to establish the parameters which determine the type of failure. Some of this work requires more complex testing facilities than are currently available but within the next decade more progress is possible in this field when new equipment becomes available and computer programs are used for the analysis of results.

8.8. DESIGN FEATURES OF STRUCTURAL ELEMENTS

On the basis of tests on prototype elements data are provided in various codes (7) and regulations (3, 4, 6) on the sizes of structural elements and their protection to meet the fire-resistance requirements. These are usually expressed as the minimum sizes for the construction, with the materials expressed in generic terms. A number of other publications provide the data on proprietary constructions where these have been tested but as yet no general design guides have been prepared which permit an analysis to be made.

P.C.S—19

8.8.1. Beams

A steel beam because of the high thermal conductivity of the material, is unlikely to provide fire resistance for any appreciable period. A rolled steel section having a weight of 30 kg/m is expected to fail as a simply supported element in 10–15 minutes time. Hence all steel beams require to be protected by an encasement. A great variety of materials is available for this purpose ranging from concrete to light density spray-on products. The thickness of the coating is related to its thermal conductivity and the duration for which protection is required. The aim is to prevent the average temperature of the lower flange exceeding 550 °C, a temperature (Fig. 8.7) at which the reduced strength may be just adequate to resist the imposed stresses. Heavier sections, because of their increased heat capacity, require less protection, the amount for a given period is proportional to P/A where P is the mean perimeter and A the area of cross-section. The former indicates the surface area available for heat transmission and the latter is a measure of heat capacity. It is occasionally more economical to protect deep beams or beams at frequent intervals by means of a suitably designed suspended ceiling rather than by individual encasement. This acts as a protective membrane and prevents excessive heat transmission into the cavity where the beams are located.

A timber beam represents a homogeneous material which on exposure is gradually consumed from the exposed faces. A study of their performance (21) in fire tests has shown that as the surface layers decompose the residue of carbon acts as an insulator and slows down the charring of wood in the section to a reasonably constant rate. With timber species having a density in the range 400–650 kg/m³ the rate of depletion from the exposed faces can be taken as 0·6 mm/min. This feature permits the fire-resistance to be computed from a knowledge of the strength characteristics of wood and the maximum fibre stresses developed by the imposed load.

Sections of laminated wood give as good a performance as solid sections of equivalent quality provided the glues used are phenolic or resorcinol based and allowance is made for the slightly increased charring at the interfaces of the laminates.

With concrete beams the limit state of instability is reached when the reinforcement in the tensile zone acquires critical temperatures, 550 °C for mild steel and 400 °C for prestressing tendons (26).

Insulation to the reinforcement is provided by the surrounding concrete, therefore the fire-resistance of such beams, is a function of the thickness of the concrete cover. Low-density concretes offer advantages in this connection by reducing the rate of heat transfer. Recent experiments (27) have shown that lightweight aggregate concrete beams require approximately 20 per cent less concrete cover for the same amount of protection. It is necessary to retain the concrete cover in position to obtain full benefits of its insulation, its fall by spalling can lead to a rapid temperature rise of the reinforcement and, in extreme cases, by exposing it, lead to premature collapse. On the basis of experimental work it has been found necessary to specify supplementary reinforcement, i.e. an additional membrane close to the exposed surfaces when the thickness of the cover, consisting of silicious aggregates, exceeds 40 mm. Lightweight aggregate concretes have been shown to be virtually free of the spalling phenomenon and do not require this type of remedial measure.

8.8.2. Floors

Floors provide a horizontal barrier to the passage of fire and they should not reach the limit state of stability during the course of a fire nor should they permit fire penetration by the formation of orifices or by the upper surface attaining excessive temperature. Floors, except those of timber, rarely fail by the formation of orifices. Thermal insulation is a property of the materials used in the construction whereas stability is determined by the behaviour of the tensile zone as for beams.

The only type of construction that could be termed a steel floor is that consisting of a steel deck with either a topping of concrete or a flooring of timber-based board. In either case, because of the exposed nature of the steel deck, the fire performance is likely to be poor; without concrete topping collapse can be expected in a few minutes. It therefore becomes necessary to provide a ceiling or some other type of soffit protection; various types of asbestos and plaster-based products can be used. For residential buildings using industrialized systems, a simple form of construction has been developed which uses lightweight steel channel sections closely spaced to act as beams, with wood particleboard flooring and a ceiling of plasterboard for a fire-resistance of half an hour.

The aim with concrete floors of either reinforced or prestressed type is to prevent the steel bars and tendons reaching critical temperatures. In the case of solid slabs it is a simple matter of ensuring that sufficient concrete cover exists; the thickness of cover necessary is less than that specified for beams as the flow of heat can occur only from one face. When reinforcement is distributed over the tensile zone, as is generally the case with prestressed constructions, the effective thickness of the cover is taken to be the mean of the cover to each tendon. Where the cover thickness is not sufficient, any deficiency can be made good by the use of an insulating coating on the soffit.

Concrete floors which do not possess a plane soffit, such as T, channel, or U-shaped constructions are more difficult to assess. They may be regarded as consisting of beam sections supporting a slab but experience has shown that the actual performance is usually better than would be expected on this basis. This is presumably due to the composite action between the ribs and the deck which ensures that structural instability is not caused when the rib reinforcement has reached critical temperatures. The present data are not sufficient to permit the performance of such constructions to be analysed in the same way as plane slabs.

Concrete floors comprising a deck supported by steel beams require adequate protection for the beams. The protection may consist of individual encasement for the beams or a suspended ceiling construction which can protect not only the beams but also the floor above. The space between the ceiling and the floor is often utilized for services including ventilating systems. It is important to appreciate that unless care is exercised perforations in the ceiling could cause failure of the protective membrane.

Timber floors have been traditional particularly for domestic building and are frequently used with industrialized systems. For a given design of floor and fire-resistance requirement, the floor beams must be large enough to resist collapse, taking into account the protection provided by the ceiling (21). It is customary to evaluate the time for which the beams can survive a direct exposure to the fire conditions and then choose a ceiling which provides the necessary additional protection. The ceiling also serves the function of preventing fire penetration to the upper side of the floor.

For domestic buildings of two-storey height a fairly light ceiling construction is adequate as the joists normally used are able to resist collapse for fifteen minutes and the ceiling is required to add another fifteen minutes or so to provide a floor of half an hour fire-resistance. Occasionally timber floors in existing buildings may be required to resist fire for up to two hours. This requires a more elaborate ceiling consisting of asbestos insulation boards with a layer of mineral wool or lightweight plaster applied to an expanded metal membrane.

8.8.3. Walls

Walls fulfil a similar function to that required of floors, i.e. resist collapse and prevent penetration of fire through orifices or by the excessive transmission of heat. In tests, non-load-bearing walls are subjected to restraint along the four edges, whereas load-bearing walls support an axial load but no restraint is applied along the vertical sides. With the application of heating there is a gradual weakening of materials, with the maximum effect on the heated side and a minimum on the other. This causes an increase in the magnitude of stresses on the less-heated section of the wall as well as an eccentricity of loading. The latter is further extenuated by the tendency of the heated side to expand and this induces buckling. The failure of such a wall is caused by the combined effects of compression and bending moments on a weakened section. It is rarely that load-bearing walls fail in direct compression. The analysis of the behaviour of load-bearing walls is therefore complex and no empirical relationships have been developed to express their performance. With non-load-bearing walls thermal insulation is usually the critical factor and computations are possible to determine the required thickness.

In addition to homogeneous constructions, consisting of brick or concrete, framed constructions with facings of a variety of materials may be used for walls. With steel framework the facing on the fire side has to prevent the temperature of the steel from reaching critical values. The total damage to the vertical members of a timber frame in a fire should not be such that the applied loads become collapse loads for the undamaged section.

For domestic-type buildings, requiring only a low fire-resistance, timber-framed constructions have become popular. These can easily be protected with layers of plasterboard, wood particleboard, or asbestos board. A number of systems such as CLASP, SCOLA, etc.,

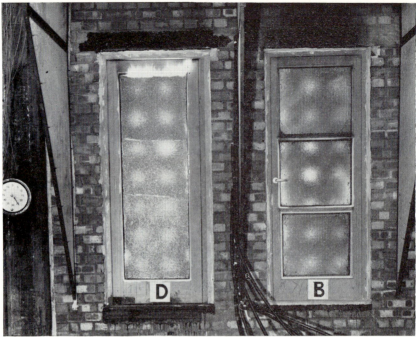

Fig. 8.12. Testing of glazed timber doors. Door D shows fire penetration near the top

have been designed where boards of one type or another are employed in the construction of the walls, and the design produced to give fire protection economically.

8.8.4. Columns

A free-standing column represents a simple case for heat transfer under fire conditions, heat transmission taking place uniformly from all faces to the inside of the section. The rise in temperature is associated with a reduction in strength of the materials and a lowering of the ultimate load capacity. As soon as this decreases to the same value as the applied load the column has reached the limit state of stability. With materials possessing a noticeable resistance to the flow of heat a temperature gradient exists across the section, the higher temperatures close to the surface causing a greater loss of strength in the outer layers. The effective size of the section is progressively reduced and the slenderness ratio increased. Hence at the limit state of stability the failure may occur either in compression or in bending depending upon the size of the original section in relation to its height. The performance of homogenous constructions can be analysed more easily than those of a composite nature.

Rolled-steel sections attain high temperatures rapidly owing to high thermal conductivity and therefore reach failure temperatures after a short exposure to fire conditions. It is possible for massive sections to delay the temperature rise to withstand heating for half an hour or so but in general this is an uneconomical method of providing fire-resistance. Hence in virtually all cases steel columns need a protection to attain fire-resistance; the nature of the materials used for this purpose determines the thickness of the encasements. Lightweight products based on asbestos and vermiculite have gained in popularity over the last decade.

Intumescent coatings have been developed which can be sprayed on to steel sections. When heated they expand by a factor of 50–100 and provide an insulating coating. As they are in the main based on plastics materials they are unable to withstand a prolonged exposure without suffering damage. Fire protection for half an hour can be provided fairly readily, for higher periods reinforcement and stabilizing additives have to be provided. Another new development for the protection of steel columns is the use of vermiculite-based materials retained around the elements by means of a light steel sheath. Such treatments can be factory applied and eliminate expensive site work. It is also possible to surround the sections on site with the sheathing and pump the insulation in the cavity.

One of the most interesting new ideas in this field is the use of water to cool the steel sections. This system is ideal for hollow sections which can be either kept charged with water or supplied with water on the occurrence of a fire. In either instance it is necessary to provide and maintain a minimum rate of flow so that the steel temperature is kept low and steam is not allowed to accumulate on the interface.

Reinforced concrete columns, by virtue of their size and the thermal properties of materials, become heated gradually and as a result show a reduction in strength; the maximum loss occurs in the outer layers. To obtain a specified performance for a given type of column it is possible to indicate the minimum dimension for the section (28). These are related to the critical section for which the imposed load becomes the ultimate load taking into account the final slenderness ratio. The usual mode of failure is in bending except for columns having a cross-section of at least 300 mm × 300 mm.

Spalling of concrete can cause the columns to fail prematurely by reducing the size of the section and permitting a more rapid heat transfer. Spalling can be reduced by the incorporation of a reinforcing membrane in the concrete cover to the vertical reinforcement for silicious aggregate concretes. Concretes made with carbonaceous aggregates or lightweight aggregate are usually free from these effects.

Columns of timber represent an exemplary case where the effect of fire is comparable to peeling off layers of the material from the outside. The charring of timber proceeds at a fairly steady rate with the undamaged section retaining most of its structural properties. The low thermal conductivity of the material ensures distinct separation between the charred and the uncharred parts. A comprehensive investigation (23) on columns of laminated timber has shown that the rate of charring is about 30 per cent faster than for beams, i.e. 0·75 mm/min as against 0·6 mm/min, with the corners becoming rounded. This permits the performance of section larger than 100 mm × 100 mm to be computed from a knowledge of the strength properties of the timber.

8.9. FUTURE TRENDS IN FIRE PROTECTION

It is obvious from the foregoing that as yet only a limited amount of progress has been made on the prediction of fire-resistance. It is the ultimate aim of workers in this field that the limit state for the structural stability of a given form of construction can be computed from a knowledge of the properties of the materials used. Progress will be made from the simpler elements such as beams and columns to the more complex walls and floors. In a number of cases empirical relationships, showing the effects of size, thickness of protective layers, and load have already been established and the next step will be the derivation of basic formulae. The progress in this field will depend on the studies being undertaken in a number of institutes on the properties of materials and on various research projects on full-size elements. It should be possible in future to predict the behaviour of whole structures under fire conditions with the use of computer programs. A start has been made in this direction by the work sponsored by the Fire Research Station at the Imperial College, London, where simple mathematical models of two- and three-dimensional structures are being examined.

Within the next ten years it is expected that some experiments will have been conducted to study the performance of three-dimensional structures and the effect of boundary conditions. These studies will use more sophisticated testing facilities than have been available so far for creating a variety of load and restraint situations for the structures. The aim is to approach the conditions which exist in complete structures with different parts providing mutual support and means for the transference of stresses. In the structural field great strides have been made in using the maximum potential of materials by design methods based on ultimate load characteristics. With the earlier design concepts the margins of safety have in practice been greater than assumed and these are being gradually reduced. With the more sophisticated ultimate load concepts it is therefore imperative to predict the behaviour of structures under fire conditions with a greater degree of accuracy.

The present trends towards industrialization of buildings will continue leading to the majority of site work comprising component assembly. Factory production methods should lead to greater uniformity of the products, the design of the joints and junctions and the site control will become more important, as the performance of the whole system may depend upon the behaviour of the joints. Special experimental facilities outlined previously will be required to provide the design data.

One of the fields likely to be explored and exploited is that of protection for steel framework. Intumescent coatings, preformed lightweight protections, factory-applied protections, sheathing of elements with light steel casing and pumping lightweight insulation into the spaces have already been mentioned. Cooling of steel sections with water as a protective measure is now being explored and further developments are likely in this connection.

In the concrete field the use of exceptionally strong materials, with compressive strengths in the range 70–140 N/mm² is being considered and reinforcing steels with strengths up to 280 N/mm² are becoming available. It is necessary to explore the properties of these materials at high temperatures, particularly the damage done to dense concretes by spalling. One of the possible methods of prevention of spalling may be by the use of glass-fibre reinforcement. It is also feasible to make concretes with exceptional tensile properties by filament reinforcement; plastics fibres are likely to be less attractive where fire exposure has to be considered but the use of carbon filaments could overcome the high temperature problems.

Wood products will find a more extensive use for housing up to four storeys in height; with laminating techniques and prefabrication fuller use will be made of its structural properties. One of the fields being explored by some laboratories is the use of wood to protect steel sections. The material is inherently capable of providing the necessary thermal insulation but the jointing techniques require to be perfected.

Plastics materials have not made great inroads in the structural field although their use in buildings is increasing at a fairly high rate. The decomposition of the present generation of plastics at elevated temperatures is the major drawback. It is possible for new composites to be developed possessing greater temperature stability. Until this occurs the majority of uses are likely to be restricted to the non-structural field.

ACKNOWLEDGEMENT

This article is Crown Copyright, reproduced by permission of the Controller, HM Stationery Office. It is contributed by permission of the Director of the Fire Research Station of the Ministry of Technology and Fire Offices' Committee.

REFERENCES

1. *Fire Research 1969* (1970). (London: HMSO).
2. MINISTRY OF WORKS (1946). 'Fire grading of buildings. Part 1. General principles and structural precautions', *Postwar Building Studies*, 20 (London: HMSO).
3. The Building Standards (Scotland) Regulations 1963. *House of Commons*. SI No. 1963: 1897 (HMSO).
4. The Building Regulations 1965. Ibid. SI No. 1965: 1373. (London: HMSO).
5. HESELDEN, A. J. M. (1968). 'Parameters determining the severity of fire. Paper 2 of Behaviour of structural steel in fire', *Ministry of Technology and Fire Offices' Committee Joint Fire Research Organization Symposium*, no. 2 (London: HMSO).
6. London Building Acts 1930–1939. Constructional By-laws (1965). London County Council, no. 4273 (London).
7. BRITISH STANDARDS INSTITUTION. 'Structural use of concrete.' Draft *Code of Practice*, no. 69/23.007.
8. ASHTON, L. A., and SMART, P. M. T. (1960). 'Sponsored fire resistance tests on structural elements', *Department of Scientific and Industrial Research and Fire Offices' Committee Joint Fire Research Organization* (London: HMSO).
9. MALHOTRA, H. L. (1966). 'Fire resistance of brick and block walls', *Fire Note*, no. 6 (London: HMSO).
10. INGBERG, S. H. (1928). 'Tests of the severity of building fires', *Natn. Fire Prot. Ass. Q.* (22), 43–61.
11. BRITISH STANDARDS INSTITUTION (1932). *Tests for the incombustibility and fire resistance of building materials and structures*, BS 476: 1932.
12. —— (1953). *Fire tests in building materials and structures*, BS 476: Part 1: 1953.
13. INTERNATIONAL STANDARDS ORGANISATION (1968). 'Fire resistance tests on structures.' ISO Recommendation R834.
14. BRITISH STANDARDS INSTITUTION (to appear). *Fire resistance tests on elements of building construction*, BS 476: Part 8.
15. ASHTON, L. A., and MALHOTRA, H. L. (1953). 'The fire resistance of prestressed concrete beams', *Joint Fire Research Organization Fire Research Note* 65/1953.
16. BUTCHER, E. G., and LAW, MARGARET (1968). 'Comparison between furnace tests and experimental fires', Paper 4 of Behaviour of structural steel in fire, *Ministry of Technology and Fire Offices' Committee Joint Fire Research Organization Symposium*, no. 2 (London: HMSO).
17. ROGOWSKI, BARBARA F. W. (to appear). 'Fire tests on system built houses', *Joint Fire Research Organization Fire Research Note*.
18. MALHOTRA, H. L. (1956). 'The effect of temperature on the compressive strength of concrete', *Magazine of Concrete Research* 8 (23), 85–94.
19. PHILLEO, R. (1958). 'Some physical properties of concrete at high temperatures.' *J. Am. Concr. Inst.* 29, 857–64.
20. LEHMAN, H., and DALZIG, G. (1960). 'Hot compressive strength of concrete', *Tonind; Ztg. u Keram. Rdsh.* 84, 414–17.
21. LAWSON, D. I., WEBSTER, C. T., and ASHTON, L. A. (1951). 'Fire endurance of laminated timber beams and floors', *National Building Studies Bulletin*, no. 13 (London: HMSO).
22. ODEEN, K. (1970). 'Fire resistance of glued laminated timber structures', Paper 2 of *Fire and structural use of timber in buildings, Ministry of Technology and Fire Offices' Committee Joint Fire Research Organization Symposium*, no. 3 (London: HMSO).

23. MALHOTRA, H. L., and ROGOWSKI, BARBARA F. W. (1970). 'Fire resistance of laminated timber columns', Paper 3, ibid.

24. *System SFI. Europefab Systems Handbook*. (1969). (London: Interbuild Prefabrication Publications Ltd).

25. LAW, M. (1969). 'Structural fire protection in the process industry', *Building* **217** (6583) 29/86–29/89; (6587) 33/65–33/68.

26. DAY, M. F., JENKINSON, E. A., and SMITH, A. I. (1960). 'Effect of elevated temperature on high tensile steel wires for prestressed concrete', *Proc. Inst. Civ. Engrs*, **16,** 55–70.

27. MALHOTRA, H. L. (1969). 'Fire resistance of structural concrete beams', *Joint Fire Research Organization Fire Research Note* 741/1969.

28. THOMAS, F. G., and WEBSTER, C. T. (1953). 'The fire resistance of reinforced concrete columns', *National Building Studies Research Paper*, no. 18 (London: HMSO).

9

Structural lightweight aggregate concrete

Adrian Pauw
Professor of Engineering,
University of Missouri

9.1. INTRODUCTION

Three significant developments have taken place in concrete technology during the past four decades which have added new dimensions to the field of concrete construction. Two of these, namely the techniques of prestressing and precasting, were primarily European developments; the third, lightweight aggregate concrete, has been, at least until recently, almost exclusively an American phenomenon.

While the process for manufacturing lightweight aggregate was developed before 1920, even as late as 1952 less than 40 000 cubic yards (30 000 m³) of this material was used for structural concrete applications. Renewed emphasis on more efficient use of materials in structures, coupled with a scarcity of high-quality natural aggregates in many regions, has led to a phenomenal increase in the use of lightweight aggregates during the past two decades. By 1965 the consumptions of these aggregates for structural concrete had increased more than hundred fold the 1952 rate.

9.2. APPLICATIONS

Initially, lightweight aggregate was primarily used in the production of masonry units. Even today, about sixty per cent of the aggregate produced annually is used for concrete block production or for insulating and fill concrete. Early applications of structural lightweight aggregate concrete were almost entirely dictated by the need for reducing the dead load. As confidence in the reliability of this material increased, and its versatility became recognized, many and varied applications were developed. In many cases it was found that architectural expression of form combined with functional design could be achieved more readily and more economically in structural lightweight concrete than in any other medium. Lightweight aggregate concrete can be used in most structural applications where normal weight concrete has been traditionally used, including multistorey building frames and floors, curtain walls, shell roofs and folded plates, bridge decks and girders, and prestressed and precast elements of all types.

Fig. 9.1. Co-op City, Bronx, New York City (1). A cooperative housing development providing 15 400 apartments in 35 high-rise buildings and 118 town houses for a resident population of 55 000. The development includes eight garages for 11 000 cars, three shopping centres and a community centre including schools for 10 000 pupils, library, fire station, theatre, post office, and places of worship. All the concrete in the superstructure of the high-rise buildings and garages, close to one million cubic yards, is expanded shale lightweight structural concrete. The buildings are of flat plate construction. In the slabs, and in the columns above the fourth floor in the high-rise buildings, 3500 lbf/in^2 (24 N/mm^2) strength is specified. Below the fourth floor the design strength is 4500 lbf/in^2 (31 N/mm^2).

Some examples illustrating the versatility of this material, both with respect to functional design and architectural expression, are shown in Figs. 9.1 to 9.3 (1, 2, 3).

The future for lightweight aggregate concrete is indeed bright. Already, in some areas such as Los Angeles and Houston, lightweight aggregate is used almost exclusively in high-rise reinforced concrete building construction. The economies to be derived by the use of lightweight aggregate in precast construction are obvious. Ever increasing form and labour costs are accelerating the development of precast concrete building systems, thus further increases in the demand for structural lightweight aggregate concrete can be anticipated.

Fig. 9.2. Forum at Inglewood, California (2). The outstanding architectural feature of this structure is a 4 ft × 8 ft (1·22 m × 2·44 m) lightweight concrete compression ring supported on 80 sculptured columns. A 407 ft (124 m) diameter radially suspended cable roofing system, attached at the centre to a steel tension ring, eliminates all columns from within the building. The columns are 57 ft (17·4 m) high with capitals flaring out to a width of 17 ft (5·2 m) to form the base of the compression ring. Each column contains 35 yd^3 (27 m^3) of structural lightweight concrete resulting in a 20-ton weight reduction over conventional concrete. The concrete was steam cured to produce a 20 hour strength of 2800 lbf/in^2 (19 N/mm^2). The 28-day compressive strength was 5000 lbf/in^2 (34 N/mm^2).

9.3. HISTORICAL DEVELOPMENT

The rotary kiln process for producing expanded shale aggregate was perfected by Stephen J. Hayde by 1917. About the same time, F. J. Straub pioneered the use of bituminous coal cinders as an aggregate for manufacture of concrete masonry units. Commercial production of lightweight expanded slag aggregate began in 1928 and the first structural quality sintered aggregate was produced in eastern Pennsylvania in 1948 using a coal-bearing shale (4).

One of the earliest structural applications of aggregate was in the construction of ships and barges by the US Emergency Fleet Building Corporation during World War I (5). Feasibility studies by J. R. Wig, a noted marine engineer, determined the need for a concrete of minimum compressive strength of 5000 lbf/in^2 (34 N/mm^2) and maximum unit weight of 110 lb/ft^3 (1760 kg/m^3). These requirements were met by the use of a rotary kiln aggregate produced in Hannibal, Missouri.

Fig. 9.3. 'Sequoias' Retirement Complex (3). A 25-storey retirement complex in San Francisco using structural lightweight concrete in slabs, beams and columns as well as in the 'Schokbeton' precast building elements and cast *in situ* architectural lightweight panels. The various required concrete strengths ranged from a low of 2500 lbf/in² (17 N/mm²) for structural toppings to a high of 5000 lbf/in² (34 N/mm²) for slabs, beams, and columns. The textured, striated panels in the low rise portion of the structure were cast in place and sandblasted.

The first use of this material was in the construction of the 7500 ton tanker *Selma*, launched in June 1919. After three years satisfactory service, the *Selma* was struck by a tug, producing a large crack near the bow. Since repair was not feasible, the ship was stripped and the hull sunk in Galveston Bay. After 34 years of exposure to sea water, the hull was inspected by divers and concrete samples removed from the hull and interior ribs. Even though in some places concrete cover

P.C.S.—20

was only $\frac{5}{8}$ in (16 mm), the reinforcement was found to be in excellent condition. Furthermore, the concrete was found to be dry at a depth of $\frac{1}{4}$ in (6 mm) from the surface, indicating negligible absorption. Compressive strength tests on 2 in (50 mm) cubes cut from the samples gave strength values ranging from 8125 lbf/in² (56 N/mm²) to 13 181 lbf/in² (91 N/mm²) (6). Examples of early applications of reinforced lightweight concrete in buildings include the Park Plaza Hotel in St. Louis and the Southwestern Bell Telephone Building in Kansas City, built during the twenties. In the early 1930s, the use of lightweight concrete for the upper roadway slab proved to be the key to the economical feasibility of the San Francisco-Oakland Bay Bridge. During World War II history repeated itself in that lightweight concrete was used in the construction of 105 ships, thereby conserving steel plate for other essential uses (7).

Shortly after World War II the development of lightweight concrete was given an impetus by a survey conducted by the National Housing Agency on the potential of lightweight concrete for home construction. Extensive studies of the properties of concrete made with lightweight aggregate were conducted by the National Bureau of Standards (8) and the US Bureau of Reclamation (9). These studies as well as earlier work by Richart and Jensen (10) and Washa and Wendt (11), focused attention on the structural potential and generated renewed interest in the use of lightweight aggregate concrete. During this same period a number of new aggregate plants were placed in operation. At first their output was used primarily for the manufacture of lightweight concrete blocks and for precast floor and roof panels. With the renewed interest generated by research studies and as experience was gained, some of this material began to be used for a few monolithic floor and roof jobs. The successful addition of four storeys to an existing department store in Cleveland, made possible by the reduced dead load without necessity of foundation modification, demonstrated potential economic applications. Similarly, in the reconstruction of the collapsed Tacoma Narrows Bridge it was found that, by the use of structural lightweight concrete in the deck, additional roadway lanes could be provided without necessity of replacing the original piers.

During the 1950s lightweight concrete was incorporated in the design of many multistorey structures to take advantage of reduced dead weight. Examples include the 42-storey Prudential Life Building in Chicago and the 18-storey Statler Hilton Hotel in Dallas. Also during this period lightweight aggregate began to be produced in Europe.

With the inclusion of provisions for structural lightweight aggregate concrete in the 1963 American Concrete Institute (ACI) building code (12) structural lightweight concrete rapidly emerged as an important sector of the structural concrete industry. Some examples of notable structures using lightweight concrete constructed in recent years include the Watergate East Apartments in Washington DC, the 60-storey Marina Towers and the 70-storey Lake Point Tower in Chicago, the TWA terminal at Kennedy International Airport and the Busch Memorial Stadium in St. Louis. Today, structural lightweight aggregate concrete is widely recognized as a versatile and economical structural material, readily available in many parts of Europe, Asia, and Australia as well as in the United States and Canada.

9.4. CLASSIFICATION OF LIGHTWEIGHT AGGREGATE CONCRETES

The mechanical properties of structural-quality lightweight aggregate concrete are very similar to those of normal-weight concrete, except for unit weight, hence design of both conventionally reinforced and pre-stressed lightweight concrete structures can be based on the same premises as those used for conventional concrete. Structural light-weight aggregate concrete, however, possesses unique properties differing in significant aspects from those of normal-weight concrete. To fully exploit the potential of this material requires careful consideration of these unique properties and their effect on structural behaviour.

Although the use of lightweight aggregate concrete has rapidly expanded and the potential for increased use in the future is un-questioned, for many engineers, architects and contractors, lightweight concrete is still a subject of confusion. This confusion is partly the result of the wide variety of natural and manufactured lightweight aggregates available for making concretes having a wide range of densities and other physical properties. Fig. 9.4 shows the spectrum of

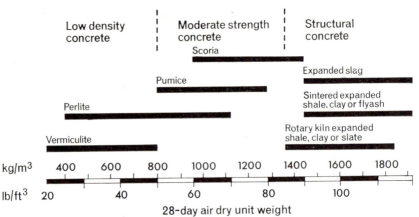

Fig. 9.4. Approximate unit weight and uses classification of lightweight aggregate concrete (4)

lightweight aggregate concretes (4). This diagram indicates the approximate unit weight range of the three types of lightweight aggregate concrete, classified according to principal function. At the low end of the spectrum are the low density concretes, weighing from 20–50 lb/ft³ (300–800 kg/m³). The concretes are of low compressive strength, ranging from about 100–1000 lbf/in² (0·7–7·0 N/mm²) and are employed chiefly for insulation purposes. The moderate strength concretes fall in the middle of the spectrum both with respect to density and strength. They are primarily used for fill where strengths ranging from about 1000–2500 lbf/in² (7·0–17 N/mm²) are adequate and where minimum weight compatible with these requirements is desirable. The structural grade lightweight concretes fall at the upper end of the spectrum and only some of the lightweight aggregates are suitable. By definition, structural lightweight aggregate concretes are concretes having a 28-day compressive strength in excess of 2500 lbf/in² (17 N/mm²) and a 28-day air-dry unit weight not exceeding 115 lb/ft³ (1850 kg/m³). Most structural quality lightweight aggregates are capable of producing concretes with a compressive strength in excess of 5000 lbf/in² (34 N/mm²) and with several of these aggregates, strengths well in excess of 6000 lbf/in² (41 N/mm²) can be achieved with reasonable cement factors. Since the weights of these structural concretes are considerably greater than those for insulating and fill concretes, insulation efficiency is lower. Nevertheless, thermal efficiency for structural lightweight concretes are substantially better than for normal-weight concrete.

9.5. LIGHTWEIGHT AGGREGATES FOR STRUCTURAL CONCRETE (4, 13, 14)

The production of structural quality lightweight concrete is predicated on the availability of suitable lightweight aggregates. Referring again to Fig. 9.4 it may be seen that several types of aggregates at the upper end of the scale are available for structural concrete. Not all of these materials, however, can be used to produce high strength concrete without the addition of natural sand and/or excessively high cement factors. The natural aggregates in this range, pumice, scoria and tuff are lightweight materials generally found in volcanic deposits. Combined with sand some of these materials can be used to produce fairly good concrete but for high strength excessive cement content may be required.

The raw materials used in the commercial production of structural lightweight aggregate are either materials found in a natural state, such as certain clays, shales and slates, or by-products from other commercial operations such as slag from blast furnaces or fly ash from the burning of coke or coal in power plants. Cinders, while used extensively for concrete block, have poor and variable concrete-making properties and are not currently used as a structural lightweight aggregate.

At the present time there are at least one hundred plants in the USA alone producing structural lightweight aggregate. Of these, approximately sixty plants employ the rotary kiln process. In this process, raw clay, shale or slate is heated and expanded under controlled conditions in rotary kilns. The other forty plants are about equally divided between sintering plants and blast furnace expanded slag plants. In the sintering process, raw clay, shale, slate or fly ash, is mixed with pulverized fuel and burned and expanded under controlled conditions on a moving grate. Expanded slag is produced by subjecting molten blast furnace slag to jets of water, steam and/or air, under controlled conditions.

In these processes expansion is produced by the formation of cells in the aggregate either by (1) formation of gases such as SO_2 or CO_2 which bloat the plastic mineral components; (2) burning off of combustible materials; or (3) formation of steam contained in the minerals. The resulting product is a lightweight cellular aggregate with cells ranging from microscopic to several millimetres in their longest dimension, dependent on the manufacturing process employed and the raw material used. For an ideal structural aggregate the resulting cell structure would be a honeycomb structure consisting of voids, moderate in size and completely separated by strong cell walls.

The output from most plants is a clinker which must be cooled, crushed and screened to produce a suitably graded aggregate. These aggregates are generally sharp, angular, and have a pitted or porous surface texture. By pre-sizing or pelletizing the raw material feed and controlling burning, to prevent or minimize agglomeration, a more rounded aggregate can be produced, both with the rotary kiln and the sintering process.

In the rotary kiln process the discharged particles are usually of semirounded shape having a hard surface or shell and such particles are commonly referred to as 'coated particles'. In some plants, some or all of this product is crushed after burning to obtain the desired gradation and the resulting product is then referred to as 'crushed particles'. The particle shape and surface texture of lightweight aggregate affect the concrete making ability in much the same way as they do for normal-weight aggregate.

It is evident from the above that the several different processes and materials available can produce many different types of aggregates ranging widely in their properties. It must be recognized, however, that all these processes and materials have been used successfully to produce lightweight aggregates with good service records and that as much or more variation is encountered in conventional aggregates now in service (4).

9.6. PROPERTIES

9.6.1. Properties of structural lightweight aggregates (4, 13)

While the properties of lightweight aggregates, as a class, can vary considerably, the physical characteristics of a lightweight aggregate from a single source are usually quite consistent—and should be expected to be so. As a class, however, lightweight aggregates possess unique properties which distinguish them from normal-weight aggregates. An understanding of these unique properties is required to exploit the full potential of these materials.

Unit weight. Structural lightweight aggregate concrete provides up to a 35 per cent weight reduction to make it a practical material in many applications where the use of normal-weight concrete would not be feasible. The finer fractions generally have a somewhat greater unit weight due to the fact that they tend to include fractions of material which have bloated least. This difference in density between aggregate fractions explains a somewhat greater tendency for segregation in stockpiles. Consistent aggregate gradation is more critical for lightweight aggregate because changes in gradation can cause fluctuation in both the unit weight and other properties of the concrete.

Maximum size. This is generally smaller than for most normal weight materials. For expanded slags and shales, the top size is usually $\frac{3}{8}$–$\frac{3}{4}$ in (10–20 mm), although some of the rotary kiln shales are available in sizes up to one inch (25 mm). With respect to such factors as workability, optimum air-content and cement content, the requirements for lightweight concrete are about the same as for normal weight concrete, when compared on the basis of the maximum size aggregate particles used in the mix. The strength ceiling, i.e. the maximum compressive strength attainable in a concrete made with a given aggregate, is influenced predominantly by the coarse aggregate. It has been found that for most lightweight aggregates the strength ceiling can be increased appreciably by reducing the maximum size of the coarse aggregate.

Particle shape. As previously noted, this can be varied, ranging from the rough and irregular crushed aggregates, with pitted and harsh surfaces, to the rounded and smooth 'coated' pebbles produced by pre-sizing the feed and controlling the burning process.

Apparent specific gravity. This is very low, compared with conventional aggregates. Since the expanded particles contain voids or dead air spaces, this property is difficult to determine, especially in the fine fraction, because of variable absorption. The specific gravity varies, as does the unit weight, with the size of the particles. Larger pieces have the lowest values while the smaller particles are heavier.

Strength. This varies from type to type. Some may be weak and friable, whereas others are tough and hard. This property need not necessarily preclude its use in structural lightweight concrete but is reflected in the range of compressive strengths for a given cement content and consistency, particularly for higher strength concretes.

Aggregate soundness. As determined by performance tests of concrete using standard freezing and thawing procedures, this is generally equal to that for good quality normal-weight aggregates. Inclusions of pop-out materials, such as burned lime or iron compounds, which contribute to unsoundness and staining, respectively, should not be permitted to be present in deleterious amounts.

Absorption. This is high compared to the 1–2 per cent water, by weight of dry aggregate, absorbed by normal-weight aggregates. Normal-weight aggregates usually contain sufficient internal moisture at the time of batching so that they absorb little if any additional water during the mixing operation. Hence in normal weight concrete the amount of mixing water required can readily be adjusted to compensate for absorption. In contrast, most lightweight aggregates can absorb 5–20 per cent water by weight of dry material. Total absorption does not normally occur during mixing and before placing hence allowance must be made for the aggregate's water demand to prevent stiffening of the mixture during the interval between mixing and placement. Thus the rate of absorption is an important factor which must be considered when uniform consistency is required in successive batches.

It should be noted that the absorbed water is not available to the cement paste in the mix during the hydration process and therefore bears no influence on the water–cement ratio. The *net* effective water–cement ratio for lightweight concrete is essentially the same, at comparable strengths, as that of normal-weight concrete.

The high-absorptive property of these aggregates, however, is not without its advantages. The absorbed water provides an internal reservoir of curing water which is available for the continued hydration of the cement, even after normal curing procedures have been discontinued. As a result, most lightweight aggregate concretes will continue to show significant gains in strength for several months after curing is discontinued.

9.6.2. Physical properties of lightweight aggregate concrete (4, 15, 16)

The summary of properties below is restricted to that part of the lightweight aggregate concrete spectrum in Fig. 9.4 considered suitable for structural concrete in load-bearing reinforced and prestressed concrete construction. With this restriction, the properties of almost all structural lightweight aggregates fall within a broad band, but with a spread not much wider than that exhibited by conventional normal-weight aggregates (17). To a somewhat greater extent than with normal-weight aggregate concrete, the properties of lightweight aggregate concrete are affected by the moisture condition of the concrete (18). Also, many of the properties appear to bear a direct functional relationship to the unit weight, e.g. lighter concretes will have a lower modulus of elasticity and lower thermal conductivity than heavier concretes of comparable strength (19). On the other hand, there is no clear line of demarcation in properties on the basis of the type of aggregate, either as a function of the raw materials or the process employed in manufacture.

Since most physical properties of concrete are related to compressive strength, an overview of the properties of structural lightweight aggregate concrete can be obtained by comparing the range of properties with those of a normal-weight concrete of equivalent compressive strength and workability (4). Such a comparison is given in Fig. 9.5. The reference values used are for an Elgin sand and gravel concrete of a compressive strength of about 4200 lbf/in² (29 N/mm²) and a 1–4 in (25–100 mm) slump. The reference concrete is approximately median among good quality normal-weight concretes with respect to most properties (15), including modulus of elasticity, drying shrinkage and creep. Such a comparison should, of course, be used with caution since many other factors affect these properties and the comparative ratios shown are not necessarily valid for other compressive strengths.

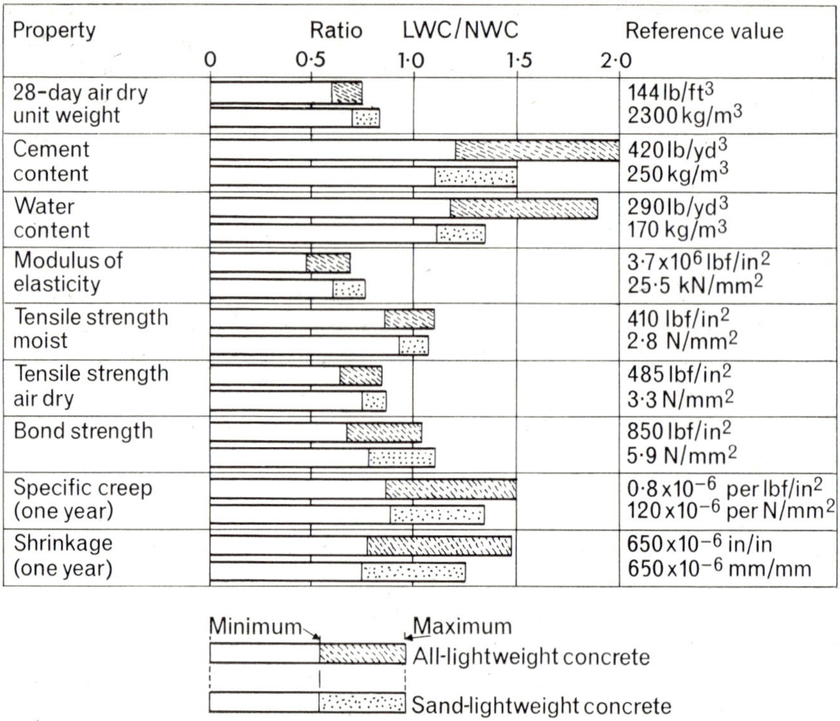

306 Structural lightweight aggregate concrete

Property	Ratio LWC/NWC	Reference value
	0 0·5 1·0 1·5 2·0	
28–day air dry unit weight		144 lb/ft^3 2300 kg/m^3
Cement content		420 lb/yd^3 250 kg/m^3
Water content		290 lb/yd^3 170 kg/m^3
Modulus of elasticity		3·7×10^6 lbf/in^2 25·5 kN/mm^2
Tensile strength moist		410 lbf/in^2 2·8 N/mm^2
Tensile strength air dry		485 lbf/in^2 3·3 N/mm^2
Bond strength		850 lbf/in^2 5·9 N/mm^2
Specific creep (one year)		0·8×10^{-6} per lbf/in^2 120×10^{-6} per N/mm^2
Shrinkage (one year)		650×10^{-6} in/in 650×10^{-6} mm/mm

Minimum — Maximum

☒ All-lightweight concrete

☒ Sand-lightweight concrete

Reference normal-weight concrete: Elgin sand and gravel, 4200 lbf/in^2 (29 N/mm^2)

Fig. 9.5. Comparative physical properties

Unit weight. Structural lightweight aggregate concrete ranges from about 85–120 lb/ft^3 (1350–1900 kg/m^3) or about 60–80 per cent that of normal-weight concrete of equivalent strength. This property is of course the principal justification for its use and can make it an economical structural material in spite of the higher cost of the lightweight aggregate (4).

Compressive strengths. Strengths up to a practical maximum of about 6000 lbf/in^2 (41 N/mm^2) can be obtained with minor increases in cement content compared with normal-weight concretes of equivalent gradation and strength. Strengths in excess of 8500 lbf/in^2 (59 N/mm^2) have been reported using certain aggregates and rather high cement contents. On the other hand, for a few aggregates the maximum strength is limited to about 5000 lbf/in² (34 N/mm²), presumably due to the lower strength of the coarse aggregate particles. With most lightweight aggregates and for a fixed cement content and consistency, replacement of the lightweight fines with natural sand increases the compressive strength. This increase is usually, but not always, accompanied by an increase in unit weight.

As with normal-weight concrete, steam curing accelerates development of compressive strength. Due to the effects of better insulation qualities of lightweight concrete, somewhat higher accelerated strengths may be obtained than with comparable normal-weight concrete cured under identical steaming conditions. Most steam-cured concretes exhibit a slight loss of compressive strength after about 90 days. This retrogression is usually attributed to microscopic shrinkage cracks which occur with air drying after the steam cure (17).

Shear (diagonal tension), tensile splitting strength and modulus of rupture. These properties are all properties closely related to the tensile strength. The tensile splitting strength can therefore be used as a convenient index of these properties. For continuously moist-cured lightweight concretes the tensile splitting strengths fall within a relatively narrow band which is not essentially different from the band for normal-weight concretes. The tensile splitting strength for air-dried normal-weight concrete is usually significantly greater than that for moist-cured concrete. Lightweight concrete specimens, however, which have undergone drying consistently exhibit much lower splitting strengths than continuously moist-cured companion specimens. This decrease appears to be due to differential shrinkage stresses resulting from a differential moisture content between the interior and exterior portions of the specimen. This differential shrinkage induces tensile stresses in the exterior shell which are balanced by compressive stresses in the interior zones and a decreased tensile splitting strength results. Sand replacement of some of the lightweight fines has been found to improve the tensile splitting strength of dried lightweight concrete. In many cases, a partial replacement of as little as one third is almost as effective as full replacement.

While the tensile strengths of moist cured specimens, both normal-weight and lightweight concrete, tend to stabilize after about a month, lightweight concrete air-dried specimens continue to gain strength for a period of six to nine months. Thus for lightweight concrete tensile strength data, 28-day air-dried specimens should be used for design values.

For design purposes the direct tensile strength may be conservatively estimated by the relationship (20).

$$f'_t = \tau \sqrt{W f'_c} \tag{1}$$

The value of τ is approximately $0\cdot33$, $(0\cdot0069)$
when f'_t = tensile strength, lbf/in², (N/mm²)
   f'_c = compressive strength, lbf/in², (N/mm²)
   W = unit weight, lb/ft³, (kg/m³)

Bond strengths. As determined by pull-out tests of deformed bars, these average about 70 per cent of the values for normal-weight concretes of comparable compressive strength. Pull-out bond strength values tend to vary over a wide range, both for normal-weight and lightweight concrete, and failure may either be due to splitting, as a result of a wedging action, or due to crushing of the concrete under the bar deformations. Sand replacement appears to be beneficial for some lightweight aggregate concretes. Further research is needed to determine the effect of the aggregate on bond strength as well as to establish the relevancy of the pull-out test as a measure of bond strength (21).

Modulus of elasticity. Normally ranges from $1\cdot4 \times 10^6$–$3\cdot0 \times 10^6$ lbf/in² ($9\cdot65$–$20\cdot7$ kN/mm²), and is therefore about half to two-thirds the value for normal-weight concrete. The modulus for both normal and lightweight concretes can be approximated by an empirical formula of the form (19):

$$E = a \sqrt{f'_c W^3} \tag{2}$$

The value of a is a function of the aggregate and ranges from about 28–38, $(0\cdot036$–$0\cdot049)$.
when E = modulus of elasticity, lbf/in², (N/mm²)
   f'_c = compressive strength, lbf/in², N/mm²)
and W = unit weight, lb/ft³, (kg/m³)
The limited test data available indicates that, for all practical

purposes, for lightweight concrete the modulus of elasticity for tension is the same as for compression.

Poisson's ratio. About the same for both normal-weight and light-weight structural concrete. A value of 0·20 is usually assumed for design purposes (17).

Creep and shrinkage. These are closely related phenomena which vary over a wide range for both normal and lightweight concrete. On the average, however, both creep and shrinkage are considerably greater for lightweight concrete. For convenience, it is generally assumed that the principle of superposition applies. Hence, creep, i.e. the dimensional change with time due to sustained stress, is usually measured by subtracting the drying shrinkage of companion unloaded specimens from the total deformation of loaded specimens. Creep, thus determined, appears to be an inverse function of the strength, with most of the creep growth taking place during the early months after the load is applied. The fact that lightweight concrete gains strength at a lower rate is therefore a partial explanation for increased creep values. Shrinkage is primarily related to the paste content and may thus increase with strength. Since, for comparable strength, the cement content of lightweight concrete is normally higher, the shrinkage can also be expected to be greater. A recently completed study (22) showed that for normal aggregate concentrations the shrinkage ratio may be approximated by the relationship

$$\frac{\epsilon_{sc}}{\epsilon_{sp}} = 1\cdot5\,(1-\sqrt[3]{V_a}), \quad V_a>0 \tag{3}$$

where: $\dfrac{\epsilon_{sc}}{\epsilon_{sp}}$ = ratio of shrinkage of concrete to shrinkage of paste,

V_a = aggregate volume concentration (volume of aggregate per unit volume of concrete).

This result is in excellent agreement with England's studies (23).

The use of sand as fines reduces both creep and shrinkage, probably through the reduction of mixing water required. Steam curing also reduces both creep and shrinkage by amounts ranging from 20–40 per cent.

P.C.S.—21

Ultimate strains. For most lightweight concretes these are somewhat greater than the value 0·003 permitted by the ACI code. The stress-strain curve for lightweight concretes tends to be linear up to higher ratios of compressive strength and as a result both the area ratio, $k_1 k_3$, and the depth ratio to the centroid of the stress block, k_2, are somewhat less than for structural normal-weight concretes. Additional research is needed to substantiate the use of the coefficients for normal-weight concrete for ultimate strength design with structural lightweight concrete.

Other physical properties. Structural lightweight concretes are surprisingly durable. Resistance to freezing and thawing has been shown to be equal to or better than that of normal-weight concrete, both with and without air entrainment. Air entrainment not only provides a high degree of durability against freeze-thaw and salt scaling but also materially improves workability. Lightweight concrete can absorb from 12–22 per cent water by volume as compared to about 12 per cent for normal-weight concrete. Any relationship which may exist between absorption and durability is uncertain and devious, as witnessed by the fact that air entrainment improves durability without appreciably altering absorption.

Cover over reinforcement is generally specified the same as for normal concrete. No evidence of any material difference in corrosion protection has been reported. The rough and hard surface characteristics of the aggregate result in good wearing qualities as testified by the excellent service record of many bridge decks constructed with lightweight aggregate concrete. Because of its lower tensile strength, however, this material is subject to 'plucking' and spalling under localized impact. A thin epoxy surfacing has been found to be a good solution for restoring and protecting areas subjected to extreme localized abrasion or wear.

The thermal coefficient of expansion is about 80 per cent that of normal-weight aggregate concrete with intermediate values resulting when sand is used as a replacement of lightweight fines. Thermal conductivity, is a function of the dry unit weight of the concrete, and ranges from a fifth to a third that of normal-weight concrete. As a result, lightweight aggregate concrete provides 20–50 per cent better fire resistance as well as improved thermal insulation (4).

9.7. DESIGN GUIDELINES

Lightweight aggregate concrete structures have been shown, both by tests of structural elements and by field performance, to behave in much the same manner as those constructed of conventional concrete. With respect to most concrete properties, the performance is merely one of degree; the basic design principles are the same and at most only minor adjustments need to be made to accommodate the effect of property differences. In the past, many successful structures have been designed using structural lightweight aggregate concrete with no other design modifications than a reduction in the dead load assumed.

For many of the properties of lightweight concrete, the difference does not warrant design modifications under usual design conditions. Thus, while the thermal coefficient of expansion is slightly lower and shrinkage is somewhat greater, modification of shrinkage and temperature reinforcement requirements is not justified. Similarly, the permeability of structural lightweight aggregate concrete and the crack width and spacing are not sufficiently different to warrant changes in minimum cover requirements over reinforcement. For other properties, such as creep and shrinkage, the dispersion is so great, both for normal and lightweight concrete, that average values can only be used as a guide for engineering judgement. When such properties are critical in determining performance, design should be based on test data for, or experience with, the specific materials used.

Other than weight, the properties of structural lightweight aggregate concrete that are significantly different to require design modifications are tensile strength and modulus of elasticity.

Flexural elements governed by flexural strength may be proportioned the same as conventional concrete beams and slabs subjected to the same total load. This procedure is justified since the ultimate strength design requirements for flexural computations apply without modification to structural lightweight aggregate concrete. The effect of lower tensile strength (Equation 1), however, should be considered in: (a) providing for shear and diagonal tension; (b) calculating the cracking load capacity of prestressed elements; and, (c) for deflection calculations, in determining the point where the section changes from a homogeneous to a cracked section. Similarly, the bond capacity may be reduced, although, bond is rarely a design criterion for high bond reinforcement (24).

When deflection criteria govern the design, minimum depths may need to be increased as much as 20 per cent to compensate for the effects of the reduced modulus of elasticity and increased shrinkage and creep. It should be noted that the decrease in flexural stiffness of the member is not directly proportional to the decrease in the elastic modulus of the concrete due to the increase in the modular ratio, i.e. the ratio of the modulus of elasticity of the steel to the modulus of the concrete. This increase in modular ratio is also beneficial, at working-load levels, in terms of distribution of stresses in the compression zone. Thus for comparable sections with equal reinforcement ratios, the neutral axis is lower in a beam section with lightweight concrete than in a beam with normal-weight concrete. As a result, concrete stresses at working-load levels are somewhat lower in lightweight concrete flexural members than in conventional concrete members of equal depth. These factors, together with reduced dead load, tend to compensate for the reduced stiffness due to decreased modulus of elasticity. Similarly the moment induced by shrinkage is comparable; the increased shrinkage potential for lightweight concrete being compensated by the lowering of the neutral axis (25).

While the lower E-value for lightweight structural concrete results in more flexible members, this reduced stiffness can at times be beneficial. In cases of impact or dynamic response, and in certain types of highly redundant structures including shells with fixed edges, the reduced stiffness tends to reduce localized stress concentrations.

When time-dependent deflections are critical, they may be estimated by the 'reduced' or effective modulus method (25). The elastic modular ratio

$$n = \frac{E_s}{E_c} \qquad [4]$$

may be computed using the value of E_c determined by Equation 2.

The effect of creep is taken into account by considering the effect of the 'reduced' modulus on the section rigidity. The effective or reduced modulus is given by

$$E_c' = \frac{E_c}{(C_c + 1)} \qquad [5]$$

where $E_c' =$ effective modulus
$\quad\quad C_c =$ creep ratio (creep strain/elastic strain)

Thus the effective modular ratio becomes

$$n' = n\,(C_c + 1) \tag{6}$$

The increase in the effective modular ratio due to creep depresses the neutral axis, resulting in a decrease in the compressive stress in the concrete, a small increase in the stress in the tensile reinforcement and an increase in the moment of inertia of the 'transformed' section. Fig. 9.6 gives the ratio of the depth, k_∞, of the compression zone after creep to the initial depth, k_0, for a simple reinforced beam in terms of the product of the parameters p and n, and the creep ratio C_c. The parameter p is the reinforcement ratio, A_s/bd. The ratio of the peak deflection y_∞, to the elastic deflection, y_0, is given in Fig. 9.7.

The effect of shrinkage on deflection may be estimated by considering the warping moment in the member due to the restraining force in the reinforcement to free shrinkage. Since shrinkage is accompanied by creep, the section rigidity should be computed on the basis of the 'reduced' modulus of elasticity of the concrete (25).

An alternate approach for estimating time-dependent deflections has been proposed by Branson, based on the use of coefficients and an 'effective' moment of inertia (20).

Creep appears to be primarily a function of the stress-strength and cement-paste ratios and, only to a lesser extent, is it dependent on the type and gradation of the aggregate. For design purposes, creep strain may be assumed to be proportional to stress. Fig. 9.8 (22) shows that for most concretes the one-year creep strain for a stress to strength ratio of 50 per cent falls within a relatively narrow band, defined by the relationship:

$$\epsilon_c = 0{\cdot}0045\,V_p\,,\ \pm 25\ \text{per cent} \tag{7}$$

where $\epsilon_c =$ one-year creep strain for $f_c = f'_c/2$
$\quad\quad V_p =$ paste ratio (volume of effective water plus cement per unit volume of concrete).

The one-year creep coefficient may, in turn, be estimated by the equation

$$C_{c1}^- = \frac{2\epsilon_c E_c}{f'_c} \tag{8}$$

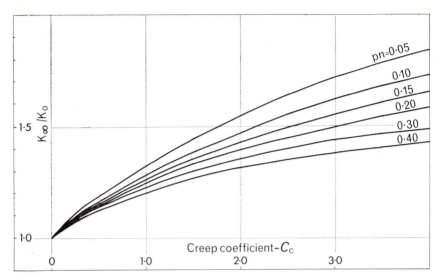

Fig. 9.6. Ratio of ultimate to instantaneous depth of compression zone

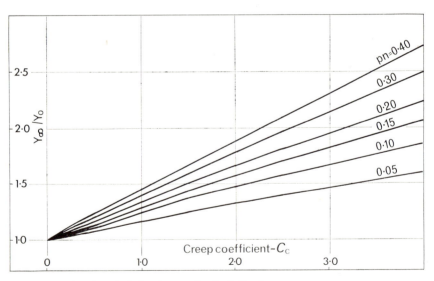

Fig. 9.7. Ratio of peak deflection to elastic deflection

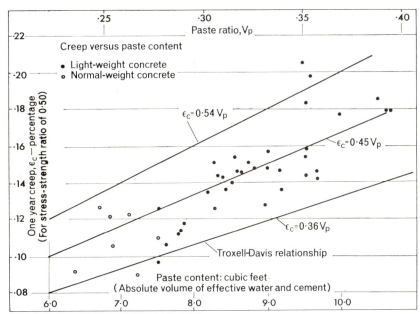

Fig. 9.8. Creep/paste ratio relationship

where C_{c1} = one-year creep coefficient
f_c' = 28-day strength of the concrete
E_c = elastic modulus of the concrete (Equation 2).

Based on an empirical relationship suggested by Branson (20), the ultimate or limiting value of the creep coefficient is approximately 30 per cent greater. The creep coefficient at any age, t days, may thus be estimated by the relationship

$$C_t = 1 \cdot 3 \, C_{c1} \frac{t^{0.60}}{10 + t^{0.60}} \qquad [9]$$

where C_t = creep coefficient, t days after application of load.

The size and shape of structural members has been shown to be of considerable importance with respect to creep and shrinkage and, to some extent, to the tensile strength of lightweight concrete. Because these properties are related to a loss of moisture and because the rates of both creep and shrinkage tend to be greater at early ages, before the concrete has gained its full strength, thin sections and sections having a small mass factor (ratio of cross-sectional area to surface area per unit length) tend to exhibit much greater creep and shrinkage as well as reduced tensile strength (26). At the present time, American design codes do not take this shape factor into account although this phenomenon has been recognized in some of the European codes and in the CEB recommendations.

Creep is also quite sensitive to the maturity of the concrete at the time of initial loading. Based on a study by Chai (27) the limiting or ultimate creep is a function of the age t_m, at time of loading. To estimate creep of concretes initially loaded at an age different than the normal 28-day curing period, the coefficient C_t in Equation 9 may be multiplied by the coefficient (22)

$$K_m = \left(\frac{28}{t_m}\right)^{0.4} \qquad [10]$$

While lightweight structural concrete may be used in prestressed concrete members, the effect of decreased modulus and lower tensile strength must be taken into consideration in computing prestress losses and in the design of end anchorages. Although the dead load deflections will tend to be balanced by the camber due to prestress, axial shortening of the member will be greater and result in greater end movements at bearing supports. The net camber of prestressed structural lightweight concrete members tends to vary somewhat more widely. Because of the greater thermal insulation offered by lightweight concrete, temperature differentials tend to be somewhat greater. Also, being more absorptive, lightweight structural concrete members are more susceptible to warping and other distortions due to differential moisture changes.

Columns can also be proportioned on the same basis, regardless of whether lightweight or normal-weight concrete is employed, provided buckling is not a design criterion. While, because of somewhat greater shrinkage and creep, the stress division of axial loads to concrete and steel is somewhat different, ultimate strength capacity, being independent of modulus of elasticity, is the same. For long columns, however, the reduced stiffness of the section must be taken into account. Insufficient evidence is available on the performance of long columns, with a slenderness ratio greater than fifteen, made of lightweight concrete. At the present time, it would seem logical to apply a factor of 0·8 to constants in load reduction formulas when lightweight aggregate concrete is used.

Because columns constitute a relatively small fraction of the total volume of concrete used in multi-storey buildings and because of the present trend toward greater column spacing coupled with smaller column size, it has become standard American practice to use very high-strength concrete in the columns and lower strength concrete in the floor systems. The use of normal-weight concrete in columns together with lightweight concrete in the floor system is both an economical and a practical solution and helps to avoid accidental use of the wrong type of concrete in the columns.

In recent years, extensive studies by Pfeifer (28), and Hognestad (29) and by Fintel and Khan (30) have demonstrated that because of the reduction in the limiting creep due to incremental load application and the redistribution of stress due to creep and shrinkage, lightweight structural concrete can be used economically in columns. A dramatic example of this potential application is the 50-storey No. 1 Shell Plaza Building in Houston. In this 714 ft (218 m) tall concrete building, column loads were applied in 50 equal weekly increments. Under these conditions, the ratio of the measured creep strain at an age of 50 weeks was found to be only about 25 per cent of the cumulative elastic strain. An analysis of Pfeifer's data (28) has revealed that for a sustained load, analysis by the reduced modulus method tends to underestimate the compressive strains by about 10 per cent, whereas analysis by the 'rate-of-creep' method (31) used by Pfeifer overestimates these strains by about the same percentage. The latter method tends to overestimate creep because it ignores the effect of creep recovery due to internal stress redistribution (22).

9.8. CONSTRUCTION PROCEDURES

High quality structural lightweight concretes, that present no particular problems in either placing or finishing, can readily be obtained by adhering to the fundamental principles of concrete mix design and control and by considering the unique properties of the aggregate. Field problems can arise if these unique properties are not taken into consideration. Most of the difficulties—as well as the potential benefits—derive from the increased absorption and lower unit weight of lightweight aggregate.

Because of variable absorption, conventional mix design procedures and control methods are not directly applicable. Satisfactory substitute procedures, however, have been developed and should be employed (32). Air entrainment is almost always desirable, not only to improve durability but also to improve workability of the mix. Maintenance of uniform and consistent gradation is somewhat more critical because of variability of unit weight with aggregate size. Due to the lighter weight of the aggregate, lightweight concrete of a given workability does not slump as much as sand and gravel concrete. These lower slump consistencies are an advantage in placing concrete on steep slopes as, for example, in the case of shell roofs.

With respect to placing and finishing, lightweight concrete presents some advantages and also some disadvantages in comparison with normal-weight concrete. The principal advantage is, of course, the reduction in weight of the material which must be handled. Forms and shores therefore can be designed for much lighter loads. The reduced weight of the concrete which must be handled requires less energy and reduces wear and handling of equipment. The principal disadvantages resulting from the reduced weight of the aggregate are a tendency toward segregation, especially when the concrete is overworked or the mix is improperly designed. While some entrained air is desirable to increase the plasticity of the mix, an excess may produce blow holes and pock marks on the surface and make the concrete difficult to finish (6).

Lightweight concrete can be placed both by pumping and by pneumatic methods. The latter has been used successfully on many shell structures. Lightweight aggregate concrete mixes for pumped concrete applications must be specially designed to prevent either pressure bleeding or 'freezing' of the concrete in the lines. 'Freezing' results from stiffening of the mix due to increased absorption of the aggregate under pressure. Some producers use vacuum absorption to assure complete saturation of the aggregates. Entrained air has been found to be desirable in some mixes because it tends to induce a 'peristaltic' wave action in the pump lines, thereby reducing the pressure required. In the last few years chemical 'pumping aids' have become available to reduce pumping friction. These aids (*polyox—*

polyethylene oxide and *natrasol*—a cellulose ether) primarily act as thickeners and lubricants and tend to increase air entrainment. Very little is known as yet about their effect on the properties of concrete. They do not appear to have any significant effect on strength except as a result of increased air content.

Excessive vibration should be avoided to prevent segregation and bleeding which, in lightweight concrete, is much more undesirable because the lighter coarse aggregate tends to float to the top while the heavier paste and the fines sink to the bottom. When this occurs the coarse aggregates can be pressed down by tamping the surface with a 'jitterbug'—a coarse screen fitted on a rigid frame—forcing the cement paste to rise to the surface.

9.9. ECONOMIC CONSIDERATIONS

The most important decision which must be made by the designer is whether structural lightweight concrete will provide an optimum solution. This decision can seldom be made on the basis of structural requirements alone. Overall system considerations must be taken into account as well as the additional cost and the availability of suitable lightweight aggregates.

Structural lightweight aggregate concrete has been most widely utilized in buildings and similar applications where the reduced dead load justifies the increased cost of the material. In general, application of structural lightweight concrete falls into one of two categories.

The first category includes structures in which the dead load constitutes a large fraction of the total load and where lightweight concrete can be specified regardless of the cost of the material. Examples of such applications include the use of lightweight concretes in ships and in the reconstruction or modification of structures using existing foundations and/or substructures and where the total load is limited.

The second category includes applications where the decision to use structural lightweight aggregate concrete must be made on the basis of economic considerations. Factors which must be considered in selecting structural lightweight concrete include: (a) reduction in the dead load, permitting shallower sections and smaller columns and footings; (b) reduction of seismic loads; (c) construction economies resulting from lighter forms, reduction of concrete handling costs, and for precast members, easier handling and erection and lower transportation costs; (d) reduced modulus of elasticity and its beneficial and adverse effects on flexibility, including increased prestress losses in tendons; (e) thermal characteristics, including the increased insulation and improved resistance to fire damage.

Structural lightweight concrete has been used successfully for floors and roofs, both *in situ* and precast, precast wall panels, bridge girders, bridge decks, and shell roofs. This material has been particularly useful in marine applications including floating structures such as ships and floating docks because the submerged weight is only about half that of conventional concrete.

Recent design innovations and current developments in materials should make economically feasible an even wider range of applications. Structural lightweight aggregate concrete decks and floors in composite with either steel stringers or precast and/or prestressed girders have proven to be extremely economical. Voided slabs and composite sections consisting of precast units and cellular concrete fills can be used effectively to increase both the rigidity and the insulating properties of the section. Other developments currently under study and which may radically affect the application of structural lightweight concrete include the use of expansive cements to compensate for increased shrinkage, and the use of chopped wire or other fibre reinforcement to improve the tensile characteristics of concrete. While these modifications would increase the cost, this increase relative to the cost of structural lightweight aggregate concrete would be considerably smaller than for conventional concrete and therefore more readily justified. There is little doubt that with its present potentialities and with future developments structural lightweight aggregate concrete will become recognized as a superior construction material.

REFERENCES

1. ANON (1970). 'Expanded shale concrete facts', *Journal, Expanded Shale Clay and Slate Institute*, **14** (3).
2. —— (1968). Ibid., **13** (3).
3. —— (1969). Ibid. **14** (1).
4. AMERICAN CONCRETE INSTITUTE COMMITTEE 213 (1967). 'Guide for structural lightweight aggregate concrete', *Journal Am. Conc. Inst., Proc* **64** (8).

5. Willson, C. (1960). *Story of the Selma*, (Washington, DC: Expanded Shale Clay and Slate Institute).

6. Pauw, A. (1955). 'Lightweight aggregates for structural concrete', *Proc. Am. Soc. of Civil Engineers*, no. 81, Separate no. 584.

7. Tuthill, L. H. (1945). 'Concrete operations in the concrete ship program', *ACI Journal, Proc.* **41** (3).

8. Kluge, R. W., Sparks, M. M., and Tuma, E. C. (1949). 'Lightweight aggregate concrete', *ACI Journal, Proc.* **45** (9).

9. Price, W. A., and Cordon, W. A. (1949). 'Tests of lightweight-aggregate concrete designed for monolithic construction', *ACI Journal, Proc.* **45** (8).

10. Richart, F. E., and Jensen, V. P. (1930). 'Construction and design features of Haydite Concrete', *ACI Journal, Proc.* **27** (2).

11. Washa, G. W., and Wendt, K. F. (1942). 'The properties of lightweight structural concrete made with Way-lite aggregate', *ACI Journal, Proc.* **38** (6).

12. American Concrete Institute Committee 318 (1967). 'Standard building code requirements for reinforced concrete (ACI 318–63)', *ACI Manual of Concrete Practice, Part 2.*

13. Pauw, A. (1968). 'Structural lightweight aggregate concrete', Theme Vb, Final Report, *Eighth Congress, IABSE, Sept. 9–14, 1968, New York.*

14. Shideler, J. J. (1961). 'Manufacture and use of lightweight aggregates for structural concrete', *PCA Devel. Bul. D40.*

15. —— (1957). 'Lightweight-aggregate concrete for structural use', *ACI Journal, Proc.* **54** (10).

16. Jones, T. R., Jr., Hirsch, T. J., and Stephenson, H. K. (1959). 'The physical properties of structural quality lightweight aggregate concrete', *Texas Transp. Inst. Report*) (Texas: A & M College).

17. Reichard, T. W. (1964). 'Creep and drying shrinkage of lightweight and normal-weight concretes', *Natl. Bur. of Stand. Monograph*, 74.

18. Hanson, J. A. (1961). 'Tensile strength and diagonal tension resistance of structural lightweight concrete', *ACI Journal, Proc.* **58** (1).

19. Pauw, A. (1960). 'Static modulus of elasticity of concrete as affected by density', *ACI Journal, Proc.* **57** (12).

20. Branson, D. E. (1970). 'Prediction of creep, shrinkage, and temperature effects in concrete structures', Draft Report for *ACI Symposium on Creep Shrinkage and Temperature Effects, New York, April 11–17, 1970.*

21. Baldwin, J. W., Jr. (1965). 'Bond of reinforcement in lightweight aggregate concrete', Preliminary Report, *61st Annual Conv., ACI.*

22. Pauw, A. *Time-dependent deformations of reinforced concrete structures*, unpublished report for Missouri State Highway Commission.

23. England, G. L. (1965). 'Method of estimating creep and shrinkage strains in concrete from properties of constituent materials', *ACI Journal, Proc.* **62** Title 62–78.

24. Gedizloglu, A. T. (1970). *Bond requirements for reinforced concrete*, MS thesis, University of Missouri—Columbia.

25. Pauw, A., and Meyers, B. L., 'Effect of creep and shrinkage on the behavior of reinforced concrete members', *ACI Special Publication* SP-9, Paper no. 6.

26. Hansen, T. C., and Mattock, A. H. (1966). 'Influence of size and shape of member on the shrinkage and creep of concrete', *ACI Journal, Proc.* **63** Title 63–10.

27. Pauw, A., and Chai, J. Wen Yu (1967). 'Creep and creep recovery for plain concrete', Missouri Cooperative Highway Research Program, *Report* 67–8.

28. PFEIFER, D. W. (1969). 'Reinforced lightweight concrete columns', *Journal, Structural Division ASCE*, **95** (1).
29. PFEIFER, D. W., and HOGNESTAD, E. (1968). 'Incremented loading of reinforced lightweight concrete columns', Final Report, *Eighth Congress IABSE, Sept. 9–14, 1968, New York*.
30. FINTEL, M., and KHAN, F. R. (1970). 'Effects of column creep and shrinkage in tall structures', Preprint, *ACI Symposium on Creep, Shrinkage and Temperature Effects, April 11–17, 1970, New York*.
31. LEONHARDT, F. (1964). *Prestressed concrete design and construction* (Berlin: Wilhelm Ernst and Son).
32. AMERICAN CONCRETE INSTITUTE, COMMITTEE 613 (1959). '*Recommended practice for selecting proportions for structural lightweight concrete* (ACI 613 A-59) (Detroit: ACI).